Microbial Problems in the Offshore Oil Industry

PROCEEDINGS OF
THE INSTITUTE OF PETROLEUM
LONDON

1981   Number 1   Performance and Testing of Gear Oils and Transmission Fluids

              Number 2   Technical Papers 1981: IP/DGMK Joint Meeting 1981

1982   Number 1   Energy Resources and Finance

              Number 2   Petroanalysis '81

              Number 3   Technical Papers 1982: IP/AFTP Joint Meeting 1982

1983   Number 1   Health and Hazards in a Changing Oil Scene

              Number 2   Technical Papers 1983

1984   Number 1   Monitoring and Maintenance of Aqueous Metal-Working Fluids: Emulsions, Synthetics and Hydraulic Fluids

              Number 2   Technical Papers 1984

1985   Number 1   Oil Loss Control in the Petroleum Industry

              Number 2   Technical Papers 1985

1987   Number 1   Microbial Problems in the Offshore Oil Industry

MICROBIAL PROBLEMS IN THE OFFSHORE OIL INDUSTRY

Proceedings of an International Conference organized by The Institute of Petroleum Microbiology Committee and held in Aberdeen in April 1986

Edited by
E. C. Hill, J. L. Shennan, R. J. Watkinson

Published on behalf of
The Institute of Petroleum, London

JOHN WILEY & SONS
Chichester · New York · Brisbane · Toronto · Singapore

Library of Congress Cataloging in Publication Data:

Microbial problems in the offshore oil industry.
    (Proceedings of the Institute of Petroleum, London;
    1987, no. 1)
    1. Petroleum — Microbiology — Congresses.   2. Offshore
oil industry — Congresses. I. Hill, E. C.
II. Shennan, J. L. III. Watkinson, R. J. IV. Institute
of Petroleum (Great Britain). Microbiology Committee.
V. Series.
QR53.5.P48M53 1987   622'338   86-24566
ISBN 0 471 91328 6

British Library Cataloguing in Publication Data:

Microbial problems in the offshore oil
    industry: proceedings of an international
    conference organised by the Institute of
    Petroleum Microbiology Committee and held
    in Aberdeen in April 1986.
    1. Petroleum — Microbiology
    I. Hill, E.C. II. Shennan, J. L.
    III. Watkinson, R. J. IV. Institute of
Petroleum *Microbiology Committee*
665'.5'01576   TP690.8
ISBN 0 471 91328 6

Printed and bound in Great Britain

622.3382'01576

Mic

LIST OF CONTRIBUTORS

BARR, M. E. J., Micran Ltd., Berryden Business Centre, Berryden Road, Aberdeen AB2 3SA

BLUNN, G., School of Biological Sciences, Portsmouth Polytechnic, King Henry Building, King Henry I Street, Portsmouth, Hants PO1 2DY

BOBOWSKI, S., BP Research Centre, Chertsey Road, Sunbury on Thames, Middx. TW16 7LN

BOCK, E., University of Hamburg, Ohnhorststr. 18, D-2000 Hamburg 52, Federal Republic of Germany

BOUSFIELD, I. J., NCIMB Ltd., Torry Research Station, P.O. Box No. 31, 135 Abbey Road, Aberdeen AB9 8DG

BRUNT, K. D., The Boots Co., plc, Chemical Marketing Department, Head Office, Thane Road, Nottingham NG2 3AA

CASTLE, J. E., University of Surrey, Guildford, Surrey GU2 5XH

CHAMBERLAIN, A. H. L., University of Surrey, Guildford, Surrey GU2 5XH

CHATELUS, C., ARBS, Section Enzymologie et Biochimie Bacterienne, CEN Cadarache, 13108 Saint Paul les Durance Cedex, France

CONNAN, J. M. R., Elf Aquitaine, Centre Micoulau - Drag - Cesed, GO - Av. President Angot 0 64018, Pau, France

COOK, P. E. Esq., City of London Polytechnic, Dept. Biological Sciences, Calcutta House, Old Castle Street, London E1 7NT

CORD-RUWISCH, R., Laboratoire de Microbiologie Orstom, Université de Provence, 3 Place Victor Hugo, 13331 Marseille Cedex 3, France

COSTERTON, J. W., The University of Calgary, Department of Biology, 2500 University Drive N. W., Calgary, Alberta, Canada T2N 1NA

LIST OF CONTRIBUTORS

CROUSIER, J., Laboratoire de Chimie des Materiaux, Université de Provence, 3 Place Victor Hugo, 13331 Marseille Cedex 3, France

CZECHOWSKI, M., ARBS, Section Enzymologie et Biochimie Bactérienne, CEN Cadarache, 13108 Saint Paul les Durance Cedex, France

DAUMAS, S., Laboratoire de Microbiologie Orstom, Université de Provence, 3 Place Victor Hugo, 13331 Marseille Cedex 3, France

DAVIES, J. M., Department of Agriculture and Fisheries for Scotland, Marine Laboratory, PO Box 101, Victoria Road, Aberdeen AB9 8DB

DAVIES, S. R. H., Institute of Offshore Engineering, Heriot-Watt University, Research Park, Riccarton, Edinburgh EH14 4AS

DOW, F. K., BP Research Centre, Chertsey Road, Sunbury on Thames, Middlesex TW16 7LN

DOWLING, N. J. E., Department of Biological Science, Florida State University, Tallahassee, Florida 32306, USA

EDLUND, A., Department of Microbiology, National Defence Research Institute (FOA 4), S-901 82 Umea, Sweden

EUVE, A., ARBS, Section Enzymologie et Biochimie Bactérienne, CEN Cadarache, 13108 Saint Paul les Durance Cedex, France

FAUQUE, G., ARBS, Section Enzymologie et Biochimie Bactérienne, CEN Cadarache, 13108 Saint Paul les Durance Cedex, France

FLETCHER, R. L., School of Biological Sciences, Portsmouth Polytechnic, King Henry Building, King Henry I Street, Portsmouth, Hants PO1 2DY

GARNER, B. J., University of Surrey, Guildford, Surrey GU2 5XH

GAYLARDE, C. C., City of London Polytechnic, Old Castle Street, London E1 7NT

GIBSON, G. R., Scottish Marine Biological Association, Dunstraffnage Marine Research Lab., P.O. Box 3, Oban, Argyll PA34 4AD

GREEN, P. N., NCIMB Ltd., Torry Research Station, P.O. Box No. 31, 135 Abbey Road, Aberdeen AB9 8DG

GREENLEY, D. E., Rohm and Haas Co., Spring House, Pennsylvania, USA 19477

GUEZENNEC, J., IFREMER Centre de Brest, BP 337, 29273 Brest Cedex, France

GUNN, N., School of Biological Sciences, Portsmouth Polytechnic, King Henry Building, King Henry I Street, Portsmouth, Hants PO1 2DY

HAACK, T. K., Rohm and Haas Co., Spring House, Pennsylvania, USA 19477

HAMILTON, W. A., University of Aberdeen, Department of Microbiology, Marischal College, Aberdeen AB9 1AS

HERBERT, B. N., Shell Research Ltd., Shell Research Centre, Sittingbourne, Kent ME9 8AG

HERBERT, R. A. Esq., University of Dundee, Department of Biological Sciences, The University, Dundee DD1 4HN

HILL, E. C., E. C. Hill & Associates, Unit M22, Cardiff Workshops, Lewis Road, East Moors, Cardiff CF1 5EG

HOLT, D. M., Micran Ltd., Berryden Business Centre, Berryden Road, Aberdeen AB2 3SA

HOLT, M. S., Shell Research Ltd., Shell Research Centre, Sittingbourne, Kent ME9 8AG

JOHNSTON, C. S., Institute of Offshore Engineering, Heriot-Watt University, Research Park, Riccarton, Edinburgh EH14 4AS

JONES, E. B. G., School of Biological Sciences, Portsmouth Polytechnic, King Henry Building, King Henry I Street, Portsmouth, Hants PO1 2DY

KEARNS, J., Hamilton Brothers Oil and Gas Ltd., Greenbank Crescent, East Tullos, Aberdeen AB1 4BG

KIERULF, C., University of Edinburgh, Microbiology Department, West Mains Road, Edinburgh EH9 3JG

LE GALL, J., ARBS, Section Enzymologie et Biochimie Bactérienne, CEN Cadarache, 13108 Saint Paul les Durance Cedex, France

LIBERT-COQUEMPOT, M. F., ARBS, Section Enzymologie et Biochimie Bactérienne, CEN Cadarache, 13108 Saint Paul les Durance Cedex, France

MASSIE, L. C., Department of Agriculture and Fisheries for Scotland, Marine Laboratory, PO Box 101, Victoria Road, Aberdeen AB9 8DB

MAXWELL, A. V., Media Supplies, Devanha House, Riverside Drive, Aberdeen AB1 2SL

MAXWELL, S., Corrosion Specialists (North Sea) Ltd., Devanha House, Riverside Drive, Aberdeen AB1 2SL

McLEAN, K. M., Corrosion Specialists (North Sea) Ltd., Devanha House, Riverside Drive, Aberdeen AB1 2SL

MOOSAVI, A. N., University of Aberdeen, Department of Microbiology, Marischal College, Aberdeen AB9 1AS

MORGAN, T. D. B., Shell Research Ltd., Thornton Research Centre, P.O. Box 1, Chester CH1 3SH

NORQVIST, A., Department of Microbiology, National Defence Research Institute (FOA 4), S-901 82 Umea, Sweden

PARKER, C. H. J. Esq., Biotechnology Centre & School of Industrial Science, Cranfield Institute of Technology, Cranfield, Bedfordshire

PARKES, R. J., Scottish Marine Biological Association, Dunstraffnage Marine Research Lab., P.O. Box 3, Oban, Argyll PA34 4AD

QUIGLEY, T. M. Esq., BP Research Centre, Chertsey Road, Sunbury on Thames, Middlesex TW16 7LN

REES, E. A., University of Dundee, Department of Biological Sciences, The University, Dundee DD1 4HN

ROBINSON, M. J. Esq., Biotechnology Centre & School of Industrial Science, Cranfield Institute of Technology, Cranfield, Bedfordshire

ROFFEY, R., Department of Microbiology, National Defence Research Institute (FOA4), S-901 82 Umea, Sweden

RUSS, M. A., University of Dundee, Department of Biological Sciences, The University, Dundee DD1 4HN

SAND, W., University of Hamburg, Ohnhorststr. 18, D-2000 Hamburg 52, Federal Republic of Germany

SANDERS, P. F., Micran Ltd., Berryden Business Centre, Berryden Road, Aberdeen AB2 3SA

SEAL, K. J. Esq., Biotechnology Centre & School of Industrial Science, Cranfield Institute of Technology, Cranfield, Bedfordshire

SEMET, R. F., Rohm and Haas Co., Spring House, Pennsylvania, USA 19477

SHAW, D. A., Rohm and Haas Co., Spring House, Pennsylvania, USA 19477

SHENNAN, J. L., BP Research Centre, Chertsey Road, Sunbury on Thames, Middlesex TW16 7LN

SIDE, J. C., Institute of Offshore Engineering, Heriot-Watt University, Research Park, Riccarton, Edinburgh EH14 4AS

SOMERVILLE, H. J., Shell UK Exploration & Production, 1 Altens Farm Road, Nigg, Aberdeen AB9 2HY

STEELE, A. D., Shell Research Ltd., Thornton Research Centre, PO Box 1, Chester CH1 3SH

STONES, A., NCIMB Ltd., Torry Research Station, P.O. Box 31, 135 Abbey Road, Aberdeen AB9 8DG

STRANGER-JOHANNESSEN, M., Center for Industrial Research, Forskningsvelen 1, P.O. Box 350, Blindern, Oslo 3, Norway

SUTHERLAND, I. W., University of Edinburgh, Microbiology Department, West Mains Road, Edinburgh EH9 3JG

TANNER, R. S., Rohm and Haas Co., Spring House, Pennsylvania, USA 19477

THERENE, M., IFREMER Centre de Brest, BP 337, 29273 Brest Cedex, France

TOCI, R., ARBS, Section Enzymologie et Biochimie Bactérienne, CEN Cadarache, 13108 Saint Paul les Durance Cedex, France

TUGHAN, C. S., University of Dundee, Department of Biological Sciences, The University, Dundee DD1 4HN

VANCE, I., BP Research Centre, Chertsey Road, Sunbury on Thames, Middlesex TW16 7LN

WATKINSON, R. J., Shell Research Ltd., Shell Research Centre, Sittingbourne, Kent ME9 8AG

WHITE, D. C., Institute for Applied Microbiology, University of Tennessee, Knoxville TN37996, USA

WOODS, D. C., School of Biological Sciences, Portsmouth Polytechnic, King Henry Building, King Henry I Street, Portsmouth, Hants PO1 2DY

POSTER CONTRIBUTIONS

Influence of Sulphate-Reducing Bacteria on Electron Flow between two Steel Electrodes
S. Daumas, R. Cord-Ruwisch, J. Crousier

Corrosion of Steel by Anaerobic Bacteria, their Products and Mixed Component Biofilms
N. J. E. Dowling, J. Guezennec, D. C. White

Marine Fouling and Corrosion Studies on 90:10 (1.5% Fe) Cupronickel Alloy
B. J. Garner, A. H. L. Chamberlain, J. E. Castle

The Role of Sulphate-Reducing Bacteria in Hydrogen Absorption by Steel
C. H. J. Parker, K. J. Seal, M. J. Robinson

Physiological Aspects of Sulphate-Reducing Bacteria involved in Anaerobic Corrosion
C. Chatelus, M. Czechowski, M. F. Libert-Coquempot, R. Toci, A. Euve, G. Fauque, J. Le Gall

Resistance of Concrete to Microbial Sulphuric Acid Corrosion
W. Sand, E. Bock

Optimisation of Bacterial Growth Media in Offshore Water Systems
A. V. Maxwell, S. Maxwell, K. M. McLean

An Assessment of Viable Count Procedures for Enumerating Sulphate-Reducing Bacteria within an Estuarine Sediment with High Rates of Sulphate-Reduction
G. R. Gibson, R. J. Parkes, R. A. Herbert

The Use of Serological Techniques for the Detection of SRB in the Oil Industry
S. Bobowski

FOREWORD

The mushrooming growth of the Offshore Oil Industry has been welcomed by oil producers, contractors, oil users — and the versatile micro-organism. Microbial fouling, spoilage and corrosion are now familiar to most technical staff in the industry, but effective avoidance and control require an understanding of how micro-organisms live and die.

Thus, the Microbiology Committee of The Institute of Petroleum organized a three-day international conference at Aberdeen from 15th to 17th April 1986 to cover the three important deleterious aspects of microbial growth, namely (a) corrosion of steel and concrete (b) biofilm formation and macro-fouling, and (c) reservoir transformations and souring. On the credit side, microbes degrade platform discharges and alleviate their environmental impact and a lecture session was devoted to this.

Other specialist lectures dealt with the selection, use and monitoring of biocides, fuel problems in supply vessels and offshore and onshore installations and a stimulating look at the possible beneficial use of microbes in oil recovery.

The papers were presented by an international panel of experts and, with one exception, have been written up for publication in this volume. No previous microbiological knowledge is really necessary although there is much in the book of interest to the specialist.

To complement the presented papers a very successful poster session was held at Aberdeen and summaries of these posters are included. A variety of practical strategies for investigation and treatment were on show.

Hence, for the first time, an industry oriented book is published as a comprehensive reference for all those in the Offshore Oil Industry who, from time to time, need to recognize microbial problems and understand why they have occurred and how best they can be controlled. Yet, much of the contents, will be of value to other industries which suffer microbial problems as the principles of corrosion, fouling and spoilage are not confined to the Offshore Oil Industry.

E. C. Hill

CONTENTS

MECHANISMS OF MICROBIAL CORROSION

W. A. Hamilton

Department of Microbiology, University of Aberdeen

Corrosion is an electrochemical event where we have both anodic and cathodic reactions (Hamilton, 1985).

$$M \rightleftharpoons M^{2+} + 2e \qquad\qquad \text{anode}$$

$$\left. \begin{array}{l} \frac{1}{2}O_2 + H_2O + 2e \rightleftharpoons 2OH^- \;) \\[2pt]) \\[2pt] 2H^+ + 2e \rightleftharpoons 2H \rightleftharpoons H_2 \quad) \end{array} \right\} \quad \text{cathode}$$

At the anode we have the dissolution of the metal and the production of electrons, and it is therefore necessary to have a cathodic reaction where these electrons are adsorbed. The more familiar cathodic reaction occurs aerobically giving rise to rust, the alternative reaction with protons as electron acceptor being possibly important under anaerobic conditions.

In microbially induced corrosion the role of micro-organisms can be direct, creating such an electrochemical cell, or it can be indirect in that it maintains a pre-existing electrochemical cell by stimulating either the cathodic or the anodic reaction. There are a number of micro-organisms and a number of mechanisms which are considered to be involved in these roles. For example, we can have growth of microbial colonies or slimes involving a whole range of possible organisms, including pseudomonads and other aerobic species. This can give rise to a situation called a concentration cell. Most commonly a differential aeration cell is created with a low concentration of oxygen, shielded underneath the slime or colony growth, as compared to the high concentration externally in the bulk environment. Under these conditions the surface in the low concentration area becomes anodic with the dissolution of metal, while the electrons react at the cathodic region with the high concentration of oxygen giving rise to hydroxide, and the ultimate formation of metal oxides and hydroxides characteristic of aerobic corrosion, or rust.

This is the commonest corrosion mechanism occurring aerobically. It is also involved in the loss of passivation of stainless steels, the passivation being dependent upon the presence of oxygen. The corrosion effect can be exacerbated where we have present, either as the sole or as a component organism, the iron oxidizing bacteria, resulting in the formation of a tubercule at the outer edges of the bacterial growth, where the oxidation of ferrous to ferric iron causes the precipitation of ferric hydroxide. This last reaction constitutes an anodic stimulation of the corrosion process.

Very often, as a result of the development of low oxygen concentrations anaerobic conditions exist within the colony or slime growth, and under these conditions one finds sulphate-reducing bacteria present. The sulphate-reducing bacteria are a principal villain when it comes to microbially induced corrosion, although not the only villain as is clear from the instances cited above. The sulphate-reducing bacteria cause significant economic damage where the total environment is anaerobic, for instance with buried pipes in clay soils; in fact the more usual situation is one in which there is the potential both for aerobic corrosion mechanisms involving concentration cells, and for anaerobic corrosion due to sulphate reducers. The normal situation with microbially induced corrosion is complex and there is likely to be more than one mechanism of corrosion operating. Although the process recognized as causing the major damage is due to the sulphate reducers, it is likely to be dependent upon other organisms to create the conditions necessary for it to occur at all. There are several mechanisms of corrosion involved with these organisms and there are a number of hypotheses and a considerable degree of argument over the mechanisms of anaerobic corrosion. The classical hypothesis (von Wolzogen Kühr & van der Vlugt, 1934) proposes an anodic reaction as before, with a cathodic reaction in which it is suggested that protons are the electron acceptor giving rise to hydrogen. Sulphate reducers will then oxidize that hydrogen to produce sulphide which will react with the iron to give ferrous sulphide as the principal corrosion product. This is an over simplification because in fact there are two reactions at the cathode; one forms atomic hydrogen from the proton and electron, followed by a combination reaction which gives molecular hydrogen. It is the molecular hydrogen which sulphate reducers can oxidize with the resulting production of sulphide. King and Miller (1971) proposed a modification of the classical hypothesis in which the ferrous sulphide corrosion product exists as a film at the metal surface. In purely electrochemical terms this ferrous sulphide film is cathodic with reference to unreacted iron so that it increases corrosion by stimulating the cathodic reaction. The ferrous sulphide film effectively increases the surface area of the cathode and so facilitates the reaction of the sulphate reducers with cathodic hydrogen. In these two ways, King & Miller (1971) have modified the classical hypotheses and have stressed the importance of sulphide as stimulating the overall reaction, but the fundamentals of the mechanism are not greatly altered. Costello (1974) on the other hand, was critical of some of the analysis and experimentation behind these ideas and suggested that it was more likely that hydrogen sulphide itself was the electron acceptor at the cathode giving rise to sulphide as HS^- and to hydrogen; note that it is molecular hydrogen which is formed directly from this reaction, which again requires to be oxidized by the sulphate-reducing bacteria.

Two other hypotheses have proposed different accents. Schaschl (1980) has suggested that elemental sulphur is the corrodent. This elemental sulphur could be produced from sulphide by autoxidation and here again note that the presence of oxygen is an important component of the mechanism of anaerobic corrosion. Schaschl suggests that there is a sulphur concentration cell, analogous to the oxygen concentration cell, by the creation of an anodic region underneath the microbial growth shielded from the higher concentration of the sulphur generated by autoxidation in the aerobic zone surrounding the cellular mass. Work done by Hardy (1983) has given some interesting information on this type of situation, by looking at the effect on corrosion, as measured by electrical resistance probe measurements, of sulphate-reducing bacteria. Growing with nitrogen anaerobically there was very little corrosion but when the reaction was sparged with air the rate of corrosion was increased very sharply. Under these conditions, Hardy noted a stimulation of corrosion from about 30.5 μm per year up to 2.25 mm per year; that is approximately a 100 fold increase in corrosion resulting from air sparging. Hardy did not suggest any mechanism but it is possible that the sparging with air gives rise to the autoxidation of sulphide to sulphur. Whatever the mechanism, these data stress again that in anaerobic corrosion oxygen appears to play a critical role.

There is a fifth hypothesis due to Iverson (1984). One of the early champions of the classical hypothesis, Iverson has recently suggested that the corrodent is in fact a volatile phosphorus compound, not yet identified but produced by the sulphate-reducing bacteria. Various phosphorus sulphide compounds have been identified as corrosion products, including vivianite (see also following paper by Moosavi & Hamilton).

Thus there are five broad hypotheses which have points of overlap and points of difference. What they have in common is the necessity for the growth and active metabolism of the sulphate reducers, and in most cases they stress the importance of hydrogen oxidation and of sulphide production although there is argument over which is the more important of these two. There has however been some recent experimental evidence looking further at the corrosion mechanism from the point of view of the growth of sulphate reducers, hydrogen oxidation and sulphide production. The first experiment of note is again from work by Hardy (Hardy & Bown, 1984). Here the authors examined the production of $[^{35}S]$-labelled sulphide from $[^{35}S]$-labelled sulphate added to their experimental system which contained a marine sulphate reducer in the presence and the absence of a steel working electrode. In the absence of this electrode there was effectively no sulphate reduction. Although the organisms are grown on hydrogen:CO_2 and acetate, here they are being incubated under nitrogen:CO_2 with acetate and these organisms cannot oxidize the acetate; they can only use it as a source of carbon and require something to oxidize, eg hydrogen, in order to supply the energy for carbon assimilation. Therefore the presence of an oxidant is required in order to obtain activity in the form of sulphate reduction under these conditions. When Hardy & Bown introduced the steel electrode, they had such an oxidant and sulphate reduction was recorded. The theory is that the steel electrode is corroding and therefore producing hydrogen at the cathode, and that this hydrogen is being oxidized by the sulphate-reducing bacteria. This is a clear demonstration that cathodic hydrogen produced electrochemically can serve as a source of metabolic energy and be oxidised by the sulphate reducers. Hardy & Bown suggested that although this was true, it was still likely that sulphide production was by far the more important aspect of corrosion. In my own laboratory, we have recently taken this further (I. P. Pankhania, A. N. Moosavi, unpublished results). Our experimental system consists of a reaction vessel sparged with nitrogen:CO_2 and containing acetate medium with hydrogen:CO_2 plus acetate-grown cells. There are reference electrode, working electrode, pH electrode, and a range of electrical control devices so that we can modify the potential on the working electrode. Initially in the absence of the working electrode there is no growth of sulphate reducers in this nitrogen:CO_2 plus acetate medium. Neither is there significant increase in hydrogenase levels, or decrease in acetate or sulphate concentrations in the medium. If the mild steel working electrode (diameter 1 cm, polished to a 600 grit finish and degreased in acetone) is then introduced but the potential held at a value (-100 mV) such that we would not expect any corrosion to take place nor hydrogen to be produced, we still get no growth, no increase in hydrogenase and effectively no sulphate being used. If the potential of the working electrode is now lowered to a corrosive value (-1400 mV) however, where hydrogen will be being produced at the electrode, we find that we get growth of the population, hydrogenase activity of the cells increases and sulphate is utilized from the medium. This is a transient phenomenon and growth, hydrogenase production and sulphate utilization cease after a period of some hours. If the working electrode is removed and at this point hydrogen:CO_2 is introduced as the gas phase, we get a rapid take-off of hydrogenase activities and growth, and a further decrease in sulphate. The cells therefore have not been killed by the presence of the corroding electrode, it is simply that the reaction has stopped and when more hydrogen is supplied then metabolic activity is readily re-established. This is a transitory effect similar to that reported by Hardy & Bown (1984) and both sets of authors have suggested that the sulphide produced during sulphate reduction and growth has poisoned the combination reaction which converts atomic to molecular hydrogen. These experiments have therefore shown that cathodic hydrogen can indeed support the growth of sulphate reducers, as required by the classical hypothesis, and in the modifications due to King & Miller and to Costello.

These experimental systems, and the whole range of studies carried out by earlier workers in this field, have had this much in common; a pure culture of a sulphate-reducing bacterium in a homogeneous suspension under regulated anaerobic conditions, no oxygen present and with a fairly simple defined medium. In the cases cited above acetate was the carbon source, but most often it has been a lactate medium that has been employed. This reflects not so much on what happens in nature, but rather what is thought about the sulphate-reducing bacteria; strict anaerobes using a restrictive number of carbon sources such as lactate and metabolically independent of other micro-organisms. Some new information has come forward about sulphate reducers which shows us that this is not representative of sulphate reducers in nature, and allows us to appreciate that our experience of corrosion in the field does not reflect this sort of situation either. From this new point of understanding we can perhaps develop new experimental methods for looking at the mechanisms of anaerobic corrosion.

The sulphate reducers were previously thought of as being a restricted range of organisms, perhaps just two genera and with a very limited range of nutrients (Postgate, 1984). Work particularly of Widdel (Pfennig, Widdel & Trüper, 1981), and since then of others, has demonstrated that there are at least nine different genera of sulphate reducers. They can be either Gram positive or Gram negative; most of them are rod shaped or vibrios, but some are lemon shaped and others filamentous; although there are those which are restricted in nutrition, we now find some which will not produce acetate as a partial end product but will in fact use acetate as a source of carbon and energy, while others can use a wide range of organic materials including benzoate and long chain fatty acids, and even some which are autotrophic and grow on CO_2; a very common property is that many sulphate reducers do indeed have the capacity to oxidize molecular hydrogen. The sulphate-reducing bacteria are a more extensive group of organisms with a much wider range of metabolic activities than we had previously appreciated, and if we always use a lactate-based medium to isolate them we are liable to get a false picture of what organisms are present in any given environment and what their potential or real activities are.

These data have been obtained from laboratory experimental studies with pure cultures, but there are also ecological studies (Sørensen et al, 1981) which have indicated that hydrogen and acetate, and to a lesser extent

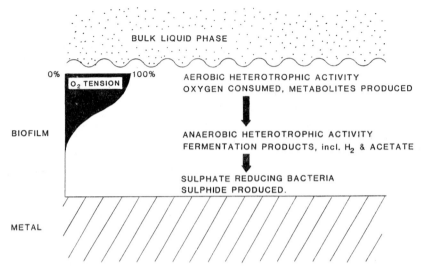

FIG 1   Model for anaerobic corrosion.

propionate and butyrate, are the principal sources of energy and carbon in most natural environments. Other ecological studies again indicate the importance of oxygen. In a marine sediment Jørgensen (1977) has shown that more than 60% of the sulphate reduction is taking place in the top 10 cm of the sediment. Thus the sulphide produced will be able to diffuse into the oxidized sediment or into the oxidized sea water above and be reoxidized by micro-organisms such as *Beggiatoa*. In this way there is no build-up of sulphide which would otherwise poison the system. At the same time, the sulphate required in the anaerobic zone diffuses from the aerobic zones. Thus the maximal activity of the sulphate reducers, although they are obligate anaerobes, occurs close to the aerobic interface. The involvement of oxygen and of the aerobic/anaerobic interface is important to rapid turn-over in such an ecological situation.

Figure 1 shows the model which we have put forward to investigate corrosion by sulphate reducers, incorporating the nutritional, physiological and ecological concepts which have been described. The model depicts a metal surface with a microbial biofilm, and at the outer surface of that biofilm the bulk phase may be a totally aerobic environment; for example, with marine fouling on the outer surface of an offshore oil platform the sea will be aerobic, while with a water- or oil-carrying pipeline the flowing liquid may or may not be aerobic depending on the conditions. Even where it is aerobic however, there will be present in the biofilm a wide range of organisms; in the case of marine fouling it will include higher organisms, and in the case of microbial growth in internal pipework it will be exclusively bacterial. The organisms present may be phototrophic or heterotrophic, aerobic or faculative, and capable of metabolizing whatever material is present in that bulk phase, for example, hydrocarbons or general organic detritus. As a result of the activities of these organisms end-products will be produced; breakdown of hydrocarbons giving rise to fatty acids, breakdown of polysaccharide materials giving sugars, and so forth. These degradative activities generally require oxygen, and at a certain thickness of biofilm the rate of diffusion of oxygen into the film will be less than the rate at which it is utilized by aerobic and facultative species so that within the biofilm anaerobic conditions will develop. Facultative and anaerobic organisms will be present and they will further degrade the fatty acids and sugars to a variety of fermentation products of which hydrogen and acetate will be key members. Thus ideal conditions will have been created for sulphate reducers. From material that they cannot metabolize such as hydrocarbons, suitable nutrients will have been produced; in what would otherwise have been an aerobic environment, anaerobic conditions will have been developed. Thus the necessary nutrient and physicochemical conditions will have been evolved to enable sulphate-reducing bacteria to live and grow, but to do so totally dependent upon the activities of other biological species present in the same biofilm, and ultimately even dependent upon the aerobic conditions which allowed these organisms to carry out their initial degradative metabolism. There is also the potential for sulphide to diffuse outwards to the aerobic regions, either through the film or where there is a patchiness in the film itself, and by chemical autoxidation give rise to sulphur, if indeed this is an important component of the mechanism. There is the potential for hydrogen produced electrochemically at the surface to act as a source of metabolic energy. The complete biofilm therefore is necessary for the metabolic activities and physicochemical conditions required for the development of the maximum potential for anaerobic corrosion by the sulphate reducing bacteria. Can we now develop methodology which allows us to test corrosion mechanisms under these conditions, and with increased knowledge either reinforce existing prejudices or perhaps come up with some new understanding.

In my own research groups, the systems we have been looking at include concrete and steel jacket production platforms with their water injection systems, oil risers, platform-to-shore pipelines, discarded drill cuttings, and marine fouling. Specifically, we have been concerned with water injection systems where microbiological problems are important from the points of view of corrosion of the system itself and the potential dangers of reservoir souring. Other areas of particular interest are the drilling legs on concrete production platforms, marine fouling, and discarded drill cuttings. The technique we employ is to put in position corrosion coupons which after a fixed period of exposure can be retrieved and subjected to corrosion and microbiological analyses. The coupons can either be unprotected or cathodically protected at -950 mV by attachment of a sacrificial zinc anode. Coupons are sited in the environment of choice for periods of a few weeks (water injection system) or up to two years (marine fouling or sea-bed sediments). An essential feature of this technique is that the critical first steps in microbiological analysis are carried out immediately the coupons are retrieved on the platform. With this system we have been analysing weight loss, corrosion products, and the numbers and types of micro-organisms present, as determined by standard counting techniques. The relevance of the numbers of bacteria that may be cultured on selective media to their *in situ* environmental activity is always open to doubt however. If our model has any validity at all then what we require is an assay of the activity of the complete undisturbed biofilm in order to gain a direct understanding of what is going on, as the organisms within the biofilm are interdependent, both in terms of nutrients and physicochemical conditions. Fortunately there is a unique activity which tells us what the sulphate reducers are doing and this we can assay directly by putting in [^{35}S]-labelled sulphate and observing its conversion to [^{35}S]-labelled sulphide. This is an exclusive assay of sulphate reducer activity and it can be carried out using a coupon with its complete and undisturbed biofilm still present. Furthermore, the assay can be carried out under conditions chosen to reflect the environment in which the coupon was exposed and so gives a rapid (24h) assay of biological activity of direct relevance to the attendant corrosion and related problems. Maxwell has established the modifications and controls necessary for the use of this radiorespirometric assay with metal corrosion coupons (Maxwell & Hamilton, 1986a).

We have applied this technology in a number of offshore environments. For example, coupons were suspended for 72 days in the flooded drill leg of a concrete production platform and the numbers and activities of organisms in the bulk phase compared with those on the metal surface of the corrosion coupons. Whereas the numbers of sulphate reducers in the bulk sea water phase were $10^3 . 1^{-1}$ on the coupons there were $10^8 . m^{-2}$, but more importantly, a [^{35}S]-sulphate reduction rate of $500 \mu moles . m^{-2} . day^{-1}$ was recorded for the coupons as compared to non-detectable levels in the bulk sea water. Although these data give us no information on any corrosion effects resulting from microbial activity, they do stress the concentration of bacterial numbers at surfaces and the likely requirement for biofilm development to create the necessary conditions for sulphate reduction activity.

Other studies have looked at the possible effects of cathodic protection on the microbial activity associated with metal test coupons (Maxwell & Hamilton, 1986b). After some 160 days exposure in a mildly polluted harbour environment, it was noted that the sulphate reduction rates on protected and unprotected coupons were 40 nmoles. coupon^{-1}. day^{-1} and 230 nmoles. coupon^{-1}. day^{-1}, respectively. The addition of glucose to the assay system raised these activities to 380 nmoles. coupon^{-1}. day^{-1} and 490 nmoles. coupon. day^{-1}. It appears therefore that in addition to minimizing corrosion *per se* cathodic protection may also be reducing microbial activity, at least as regards the sulphate-reducing bacteria. The very considerable stimulation given to both protected and unprotected steels by the addition of glucose however, suggests that the rate-determining factor may be substrate limitation. In this regard it is important to point out that glucose is not itself a substrate for sulphate-reducing bacteria and that this finding reinforces the concept that these bacteria are critically dependent upon other species of micro-organisms within the biofilm. Our view of cathodic protection and sulphate-reducing activity is that any effect is likely to be indirect in that the formation of corrosion products and sulphide films helps to generate at the metal surface the anaerobic conditions necessary for the growth and activity of the sulphate-reducing bacteria. This therefore adds to the suggestion of King & Miller (1971) that the sulphide film is itself directly involved in the electrochemical corrosion reaction.

Longer exposures of up to two years under marine fouling on offshore production platforms in general do not show a similar effect of cathodic protection on sulphate reduction. The range of activities recorded with unprotected steel was 30-200 nmoles. 10cm^{-2}. day^{-1} and with protected samples, 0–150 nmoles. 10cm^{-2}. day^{-1} (Sanders & Hamilton, 1986). One explanation of these contradictory findings might be that over the longer time periods the anodes are coming to the end of their useful life and that at the time when the sulphate reduction activity is being assayed the effective cathodic protection is greatly reduced. It is more likely, however, that we are trying to establish an overall picture on the basis of a restricted number of single snapshots, and that only when we have more extensive data acquired regularly at a single site over an extended period will a less equivocal picture emerge.

We are now engaged in such an extended study, with coupons buried in sea-bed deposits of discarded drill cuttings. We have already noted (Sanders & Hamilton, 1986) high rates of sulphate reduction associated with cuttings from diesel-based muds (100–5000 nmoles. g^{-1} day^{-1}) as compared to water-based muds (100–200 nmoles. g^{-1}. day^{-1}) or natural sediment from a polluted estuary (100–400 nmoles. g^{-1}. day^{-1}). These activities are matched by the numbers of sulphate-reducing and hydrocarbon-oxidizing bacteria that can be cultured from the various deposits. When corrosion coupons are introduced into cutting deposits similar patterns of activity and numbers are noted. With diesel-based muds the figures are 500 nmoles. coupon^{-1}. day$^-$, 10^4 sulphate reducers.

coupon^{-1} and 10^{-2} hydrocarbon degraders. coupon^{-1}; with low toxicity muds the figures are 400, 10^3 and 10^1, and with water-based muds, 300, 10^2 and 10^1. The corrosion rates in these three environments are, respectively, 60, 50 and 30 mg. dm^{-2}. day^{-1} (mdd). Where cathodic protection is applied these figures reduce to 2, 0.5 and 0.5 mdd. The most interesting statistic, however, relates to the anode life. In each case the anodes had a design-life of 8 years. Although the anodes remained active in low toxicity and water-based drill cuttings for longer than three years (duration of study, so far), those in the diesel cuttings were exhausted in less than two years. That is to say, although cathodic protection reduces corrosion even in the presence of high sulphate reduction activity, under these conditions there is excessive pressure on the anode leading to reduced life and ultimately increased corrosion risk.

Although such studies in the offshore environment have direct relevance to the problems experienced by the oil industry, they are inevitably limited to terms of their scientific rigour. We have therefore mounted a parallel series of laboratory experiments designed to give less equivocal data on the actual mechanisms of anaerobic microbial corrosion.

REFERENCES

Costello, J. A. (1974). 'Cathodic depolarisation by sulphate-reducing bacteria', *S. A. J. Sci.*, **70**, 202–240.

Hamilton, W. A. (1985). 'Sulphate-reducing bacteria and anaerobic corrosion', *Ann. Rev. Microbiol.*, **39**, 195–217.

Hardy, J. A. (1983). 'Utilisation of cathodic hydrogen by sulphate-reducing bacteria', *Br. Corros. J.*, **18**, 190–193.

Hardy, J. A. and Bown, J. L. (1984). 'Sulphate-reducing bacteria: their contribution to the corrosion process', *Corrosion*, **40** (12), 650–654.

Iverson, W. P. (1984). 'Mechanism of anaerobic corrosion of steel by sulfate reducing bacteria', *Mat. Perform.*, **23** (3), 28–30.

Jørgensen, B. B. (1977). 'The sulphur cycle of a coastal marine sediment (Limfjorden, Denmark)', *Limnol. Oceanogr.*, **22**, 814–832.

King, R. A. and Miller, J. D. A. (1971). 'Corrosion by the sulphate-reducing bacteria', *Nature*, **233**, 491–492.

Maxwell, S. and Hamilton, W. A. (1986a). 'Modified radiorespirometric assay for determining the sulfate reduction activity of biofilms on metal surfaces' *J. Microbiol. Meth.*, (in press).

Maxwell, S. and Hamilton, W. A. (1986b) 'Activity of sulphate-reducing bacteria on metal surfaces in an oilfield situation', in *Biologically Induced Corrosion* (Ed. S. C. Dexter), in press, National Association of Corrosion Engineers, Houston.

Pfennig, N., Widdel, F. and Trüper, H. G. (1981). 'The dissimilatory sulfate-reducing bacteria', in *The Prokaryotes* (Eds. M. P. Starr, H. Stolp, H. G. Trüper, A. Balows, H. G. Schlegel), pp 926–940, Springer-Verlag, Berlin.

Postgate, J. R. (1984). *The Sulphate-Reducing Bacteria*, 2 ed, Cambridge University Press, Cambridge.

Sanders, P. F. and Hamilton, W. A. (1986). 'Biological and corrosion activities of sulphate-reducing bacteria in industrial process plant', in *Biologically Induced Corrosion* (Ed. S. C. Dexter), in press, National Association of Corrosion Engineers, Houston.

Schaschl, E. (1980). 'Elemental sulphur as a corrodent in deaerated, neutral aqueous solutions', *Mat. Perform.*, **19** (7), 9–12.

Sørensen, J., Christensen, D. and Jørgensen, B. B. (1981) 'Volatile fatty acids and hydrogen as substrates for sulfate-reducing bacteria in anaerobic marine sediment', *Appl. Environ. Microbiol.*, **42**, 5–11.

von Wolzogen Kühr, C. A. H. and van der Vlugt, L. S. (1934) 'The graphitization of cast iron as an electrobiochemical process in anaerobic soils', *Water* (den Haag), **18**, 147–165.

MICROBIAL CORROSION STUDIES IN A MARINE SULPHURETUM

A. N. Moosavi & W. A. Hamilton

Dept. Microbiology, University of Aberdeen, Scotland.

INTRODUCTION

A sulphuretum is a system in which part or whole of the sulphur cycle takes place. One of the earliest papers which mentions a sulphuretum is that by Gaines (1910) in which he describes a marsh where both aerobic and anaerobic bacteria were active. As a result of their activity a sulphate — sulphur — sulphate cycle was induced in the marsh. Moreover, Gaines found high rates of corrosion of iron and steel in such a sulphuretum.

It is a well-known fact that cases of microbially induced corrosion (MIC) reported in the field are frequently more severe than those simulated in the laboratory. One likely explanation for this discrepancy is that most of the laboratory experiments on MIC are designed to study the effects of specific genera of bacteria or in some cases, a limited combination of micro-organisms. In actual field conditions, however, MIC is most often caused as a result of a number of micro-organisms acting in concert (Hamilton, 1985).

One example of a naturally occurring sulphuretum is an oil storage tank. Two genera of bacteria which have been shown to co-exist in an oil storage tank are sulphate-reducing bacteria (SRB) and the aerobic hydrocarbon oxidizing bacteria (Wilkinson, 1983, Gilbert et al 1984). This combination of bacteria has also been identified in oil reservoirs (Herbert et al, 1985) and in drill mud cuttings piles around offshore platform legs (Sanders and Tibbets, 1986). In such systems the hydrocarbon oxidizing bacteria use up the available oxygen thus providing anaerobic environments for growth of SRB. Additionally the products of the oil degraded by these bacteria serve as nutrient for the sulphate reducers (Ross, 1983).

SRB can be present in sediments piled up in stagnant areas at the bottom of oil storage tanks. Indeed, large numbers of these bacteria have been detected in oil storage cells (Lunden and Stastny, 1985). There are also other ways by which SRB can find their way into a storage tank. Figure 1 shows in very simplified form a typical oil storage operation. Initially the storage tank is filled with sea water, which will be displaced by crude oil when the tank is used for oil storage. The skimmer tank is used to separate oil droplets from the sea water expelled from the storage tank as crude

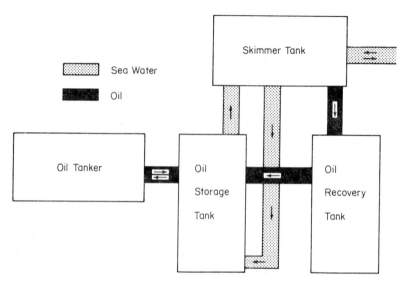

FIG 1   A water displacement oil storage system.

oil is pumped into the tank. In addition, the skimmer tank provides a source of sea water for the tank cleaning operations and a settling basin for the sludge removed from the tank. The oil skimmed off in the skimmer tank is recovered at the oil recovery tank and is then pumped back into the oil storage tank. Thus a potentially major breeding ground for SRB can be the sludge collected at the skimmer tank. These bacteria can, by the cyclic nature of the storage tank operation, find their way into the storage tank. The sea water will provide the necessary sulphate. Once SRB are established in a tank or a reservoir, it can be difficult to 'turn-off' their activity by cutting off their nutrient supply (Herbert *et al*, 1985).

SRB, by producing H_2S, can pose a problem in the oil industry in a number of ways. One is from the personnel safety point of view since H_2S is highly toxic. Additionally H_2S can result in the souring of crude oil and it can also cause corrosion of iron and steel structures. Even reinforced concrete is not immune from corrosion by SRB (Moosavi *et al*, 1985). The mechanisms by which SRB cause the corrosion of iron and steel are not yet fully understood and there are many different theories which try to explain the corrosion mechanism. These have been comprehensively reviewed by Hamilton (1985). Most hypotheses single out the biogenic sulphide as the originator of the corrosion process. An interesting theory, however, is that by Iverson (1981) which suggests that a highly corrosive phosphorus compound is involved in the corrosion process. Other interesting findings include those by Wilkinson (1983) who found that SRB grow much faster and produce more sulphide in the presence of oil; Hardy and Bown (1984) who observed that the corrosion rate of steel in the presence of active SRB increased when a limited supply of air was introduced into the system; and Videla (1985) who has concluded that the action of biogenic sulphides in the corrosion process would be enhanced through the formation of microbial consortia within biofilms on the metal surface.

One of the possible products in a sulphuretum is elemental sulphur. This can also play an important part in the corrosion process. Maldonado-Zagal and Boden (1982) have shown that elemental sulphur reacts with water over a wide temperature range to give as products SO_4^{2-}, HS^- and H^+ ions which in turn can lead to rapid corrosion of mild steel. Schaschl (1980) observed that elemental sulphur acts as a corrodent in deaerated, neutral pH aqueous solutions by forming a sulphur concentration cell, the solubility of the sulphur being the key to its activity as a corrodent. He has further claimed that the role of bacteria in the corrosion process is solely to provide a sheltering action so that the sulphur concentration cell can operate.

In the present work a marine sulphuretum containing sediment, sea water, crude oil and various bacteria was set up to study the corrosion processes which may occur in a natural sulphuretum such as an oil storage tank.

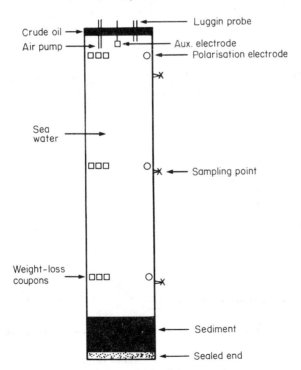

FIG 2   Laboratory model 'sulphuretum' apparatus.

EXPERIMENTAL WORK

Construction and Operation of the Sulphuretum

A perspex cylindrical tube (2 m × 7 cm), sealed at one end was used as the sulphuretum body. Three sampling ports were embodied close to the top, middle and bottom of the sulphuretum (Figure 2). The sampling ports were also made of perspex and were connected to glass taps by non-permeable rubber tubing. Hose clips were used on the tubing to regulate the water pressure. The bottom 10 cm of the sulphuretum was then covered with black sediment smelling of hydrogen sulphide, collected from the River Don near a sewage outlet. The vessel was then filled with sea water from the North Sea. Mild steel weight-loss coupons (13.5 cm^2) and polarization electrodes (2 × 1 cm mild steel cylinders, embedded in resin, polished to 600 grit and cleaned with acetone) were placed near the sampling ports. A luggin probe (connected to a calomel electrode), a platinum electrode and an air pump were inserted at the top of the sulphuretum. Also added to the top outlet was 150 ml of crude oil (2% of total vol.). At the start of the experiment a mixed culture of hydrocarbon oxidizing bacteria consisting of *Pseudomonas* and *Micrococcus sp.* capable of growing on both alkanes and aromatics was added to the oil layer. This mixed enrichment was grown on Bushnell-Haas medium (Difco) plus 1% Brent crude oil. A number of species of *Pseudomonas* were present in the enrichment. These were rod shaped, motile, gram negative, white, yellow and brown colours on agar plates. The gram positive cocci were found singly, in pairs or tetrads. They were 2–3 μm in size and can grow on hexadecane, naphthalene and anthracene.

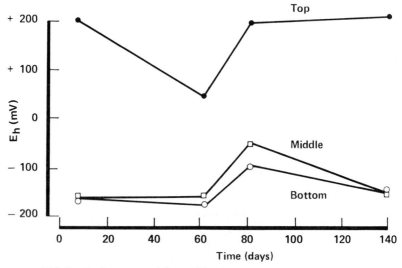

FIG 3   Redox potentials within the model 'sulphuretum'.

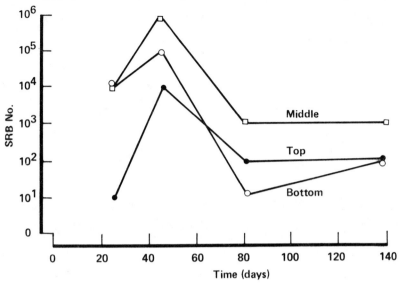

FIG 4   Distribution of SRB within the model 'sulphuretum'.

Corrosion Rate Measurements

Corrosion rates were measured by means of both weight-loss and polariz-ation techniques. Weight-loss coupons were weighed at the start and the end of the experiment while polarization measurements were carried out throughout the experiment. Corrosion rates were calculated from the polarization data using Faraday's laws. A reasonably good agreement was found between corrosion rates obtained by weight-loss and those calculated from polarization data (Table 1).

Table 1   The average corrosion rates obtained from weight-loss coupons and calculated from polarization electrodes in the sulphuretum

Electrode position	Actual corrosion rate (m.p.y)	Calculated corrosion rate (m.p.y)
Top	3.36	4.99
Middle	1.79	1.88
Bottom	1.63	1.69

Biological and physico-chemical monitoring

Samples were taken from the three sampling ports on a regular basis for measuring the pH, redox potential (Figure 3), oxygen concentration and SRB enumeration (Figure 4). The media used for enumerating SRB by MPN was Postgate's media B (Postgate, 1984).

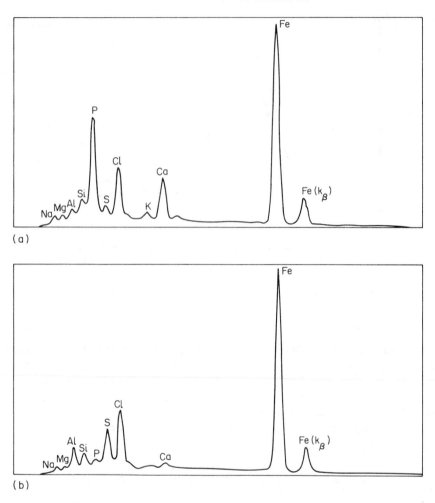

FIG 5(a)   25 kV X-ray spectrum from corrosion product labelled 'top' — large area scan on material smeared on to double-sided tape on a carbon support. (b) 25 kV X-ray spectrum from corrosion product labelled 'bottom' — large area scan on material smeared on to double-sided tape on a carbon support.

X-Ray Diffraction and Scanning Electron Microscopy

Corrosion products from weight-loss coupons at the bottom and the top of the sulphuretum were analysed by X-ray diffraction (Figure 5) while weight-loss coupons at all three points in the sulphuretum were examined by scanning electron microscopy (figures 6–8), having first been cleaned in Clark's solution and degreased with acetone.

RESULTS AND DISCUSSION

The pH values of the water taken from the three sampling ports in the sulphuretum were around the neutral value for the duration of this work. The fact that the pH at the top remained neutral is slightly surprising since during oil degradation the pH usually drops to about 5 (Reisfeld *et al* 1972). The neutral pH at the top may be accounted for by the large volume of sulphuretum which will have had a diluting effect on the water samples. The neutral pH is, however, ideal for growth of SRB.

The redox potential at the top remained positive throughout the experiment while those at the middle and bottom of the sulphuretum stayed negative. All the redox potentials decreased initially, then increased to a peak value only to decrease again (Figure 3). A comparison between these results and those from SRB enumerations (Figure 4) shows that, in general, an increase in SRB numbers coincided with a shift in the Eh towards more negative values while any decrease in SRB numbers corresponds to shifts in Eh towards more positive values. It is generally accepted that in order for the SRB to grow the redox potential of the environment should be negative (the figure of -100 mV is a common criterion). Once SRB start to grow they further reduce the Eh by producing sulphide. The initial negative Eh is usually attainable in anaerobic environments such as stagnant sediments.

The presence of SRB throughout the sulphuretum suggested that the hydrocarbon oxidising bacteria had degraded the crude oil at the top thereby providing nutrient for the SRB which were present in the sediment at the bottom. The SRB numbers initially increased in all parts of the sulphuretum (Figure 4) and then decreased with time to reach a steady value. The redox potential values at the middle and bottom of the sulphuretum suited the growth of SRB. At the top of the sulphuretum, however, where redox potentials were positive and the oxygen concentration was over 1 ppm, high numbers of SRB were nonetheless detected. This seems to indicate the SRB in the sulphuretum could withstand the presence of a limited amount of air in the system and still remain viable.

Corrosion rate data (Table 1) show that the steel coupons at the top of the sulphuretum corroded twice as much as those in the other parts. The main differences between the physicochemical conditions at the top and the rest of the sulphuretum were that the top was the most aerobic part and was in close contact with crude oil. The high corrosion rates at the top seem to be consistent with the work of Hardy and Bown (1984) who found that with air present the corrosion of steel by SRB was much greater than that found in anaerobic conditions. The fact that oxygen

FIG 6(a)   SEM of a coupon from the 'top' of the sulphuretum.

was continuously being pumped into the sulphuretum from the top might therefore be a major contributing factor to the higher corrosion rates at the top.

The X-ray spectra of the corrosion product collected from the coupons at the bottom of the sulphuretum (Figure 5(a)) revealed large amounts of phosphorus and negligible amounts of sulphur. The spectrum of the product of the top electrode (Figure 5(b)) showed negligible amount of phosphorus and a notable amount of sulphur. Thus while the corrosion product at the top seemed to be mainly iron sulphide, that found at the

FIG 6(b).

bottom appeared to be a compound of iron and phosphorus. This latter finding is in agreement with Iverson (1981) while the former is in accordance with most of the other published works in this field. The reason for this discrepancy is not clear and cannot be adequately explained on the basis of this work alone.

The Scanning Electron Microscopy (SEM) pictures (Figures 6–8) revealed that the corrosion on the top coupons was much more local in nature than that free from corrosion while the corroded areas exhibited pits much deeper than those seen on the other coupons.

FIG 7(a)   SEM of a coupon from the middle of the sulphuretum.

CONCLUSION

Preliminary findings indicate that two distinctly different corrosion mechanisms existed in the sulphuretum:

1. High corrosion rates with 'producing' Fes and pitting, but occurring in aerobic zones (cf Hardy & Bown, 1984).
2. A second corrosion mechanism in anaerobic regions consistent with Iverson (1981) where phosphorus is present in the corrosion

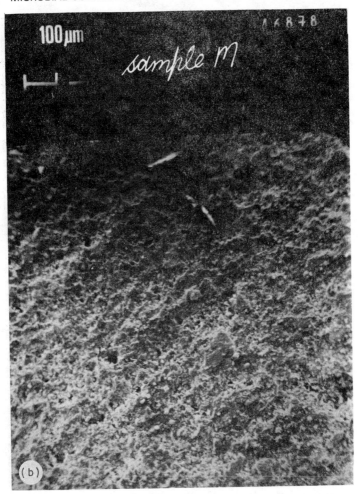

FIG 7(b).

product. This mode of corrosion was more general than the one mentioned above.

There clearly exists a potential for clarification of the above findings by using a physico-chemically and biologically heterogenous system. Research is now in progress at this Department to try to elucidate and clarify these findings.

ACKNOWLEDGEMENTS

The authors wish to thank Mr R. S. Pirrie for constructing the sulphuretum, Mr W. J. McHardy of The Macaulay Institute for Soil Research for carrying

FIG 8(a)   SEM of a coupon from the bottom of the sulphuretum.

out the SEM work and Dr J. F. D. Stott of CAPCIS for his useful comments. We also gratefully acknowledge the financial assistance from the Marine Technology Directorate of the Science and Engineering Research Council.

REFERENCES

Gaines, R. H. (1910). 'Bacterial activity as a corrosive influence in the soil', *J. Ind. Eng. Chem.*, **2**, 128–130.

Gilbert, P. D., Steele, A. D., Morgan, T. D. B. and Herbert, B. N. (1983). 'Concrete corrosion', in *Microbial problems and corrosion in oil and oil product storage*, Institute of Petroleum.

FIG 8(b).

Hamilton, W. A. (1985). 'Sulphate-reducing bacteria and anaerobic corrosion', *Ann. Rev. Microbiol.*, **39**, 195–217.

Hardy, J. A. and Bown, J. L. (1984). 'Sulphate-reducing bacteria: Their contribution to the corrosion process', *Corrosion*, **40**, 12, 650–654.

Herbert, B. N., Gilbert, P. D., Stockdale, H. and Watkinson, R. J. (1985). 'Factors controlling the activity of sulphate-reducing bacteria in reservoirs during water injection', presented at Offshore Europe, Aberdeen.

Iverson, W. P. (1981). 'An overview of the anaerobic corrosion of underground metallic structures, evidence for a new mechanism', ASTM STP 741, American Society for Testing Materials, 33–52.

King, R. A. and Miller, J. D. A. (1971). 'Corrosion by the sulphate-reducing bacteria', *Nature*, **233**, 491–492.

Lunden, K. C. and Stastny, T. M. (1985). 'Sulphate-reducing bacteria in oil and gas production', in *Corrosion '85*, NACE.

Maldonado-Zagal, S. B. and Boden, P. J. (1982). 'Hydrolysis of elemental sulphur in water and its effect on the corrosion of mild steel', *Br. Corros. J.*, **17**, 3, 116–120.

Moosavi, A. N., Dawson, J. L., Houghton, C. H. and King, R. A. (1985). 'Effect of sulphate-reducing bacteria on the corrosion of reinforced concrete', in proceedings of the International Conference on Biologically Induced Corrosion, Gaithursburg.

Postgate, J. R. (1984). *The Sulphate-Reducing Bacteria*. Cambridge University Press. Second Edition.

Reisfeld, A., Rosenberg, E. and Gutnick, D. (1972). 'Microbial degradation of crude oil: factors affecting the dispersion in sea water by mixed and pure cultures', *Appl. Microbiol.*, **24**, 3, 363–368.

Ross, D. (1983). 'Ecological studies on sulphate-reducing bacteria in offshore oil storage systems', Ph.D. thesis. University of Aberdeen.

Sanders, P. F. and Tibbets, P. J. C. (1986). 'The effects of discarded drill mud cuttings on microbial populations', in proceedings of the Royal Society meeting on Environmental effects of North Sea oil and gas developments.

Schaschl, E. (1980). 'Elemental sulphur as a corrodent in deaerated, neutral aqueous solutions', *Mat. Perf.*, **19**, 7, 9–12.

Videla, H. A. (1986). 'Corrosion of mild steel induced by sulphate-reducing bacteria. A study of passivity breakdown by biogenic sulphides.' *Corrosion*, (to be published).

Wilkinson, T. G. (1983). 'Offshore Monitoring', in *Microbial Corrosion*, The Metals Society, (1983).

FACILITATION OF CORROSION OF STAINLESS STEEL EXPOSED TO AEROBIC SEAWATER BY MICROBIAL BIOFILMS CONTAINING BOTH FACULTATIVE AND ABSOLUTE ANAEROBES

Nicholas J. E. Dowling (a), Jean Guezennec (b), and David C. White (c)

(a) Department of Biological Science, Florida State University, Tallahassee, FL 32306, USA.
(b) IFREMER Centre de Brest, BP 337, Brest Cedex, France.
(c) Institute for Applied Microbiology, University of Tennessee, Knoxville TN 37996, USA

ABSTRACT

There is increasing evidence that corrosion of metals exposed to seawater is facilitated by the presence of microbes and their products. Microbes of differing physiological types when acting in consortia appear to be more destructive than monocultures. Methods for examining consortia based on the detection of lipid biomarkers after extraction and analysis by gas chromatography/mass spectrometry (GC/MS) that are characteristic for different classes of microbes make it possible to correlate the effects of shifts in the microbial community structure to the facilitation of corrosion. This study will present preliminary evidence that a consortia of bacteria made up of a heterotrophic facultatitively anaerobic *Vibrio natriegens* and the obligately anaerobic sulfate-reducing *Desulfobacter* form a biofilm on stainless steel surfaces exposed to aerobic seawater that facilitates corrosion.

INTRODUCTION

The actuality of microbial facilitation of corrosion (MFC) of metal surfaces has been a difficult concept for some corrosion engineers (Tatnall, 1981). Often these engineers have had little experience with microbiology and the manifold metabolic capacities that these organisms possess. Part of the problem is the difficulty in routinely identifying specific microbes that may be involved in MFC. Often these microbes are exceedingly difficult to isolate in pure culture and identify (as for example the iron-oxidizing *Gallionella* spp.).

A second and possibly more difficult problem is that these microbes often are closely associated in consortia containing bacteria of multiple physiological types. These associations greatly potentiate the enzymatic versatility and thus the metabolic potential for MFC. The best studied of the microbial associations are the consortia involved in the anaerobic fermentation of complex polysaccharides (Wolin, 1979). In these assoc- iations the hydrolysis of the complex polysaccharides by one group of organisms is coupled to the conversion of the carbohydrate monomers to short chain acids and hydrogen by another group. The metabolism of the monomer utilizing fermenters requires that the partial pressure of hydrogen be less 10^{-4} atmospheres. This is possible because of another group of bacteria that utilize the hydrogen to form reduced sulphur or hydrogen.

The problems of examining the metabolic activity of consortia of bacteria of mixed physiological types required the isolation of each component of the consortia and the study of the mixed isolates. Often the organisms proved extremely difficult to culture in isolation.

Our laboratory has been involved in the development of assays to define microbial consortia in which the bias of cultural selection of the classical plate count is eliminated. Since the total community is examined in these procedures without the necessity of removing the microbes from surfaces, any microstructure of multi-species consortia is preserved. The method involves the measurement of biochemical properties of the cells and their extracellular products. Those components generally distributed in cells are utilized as measures of biomass. Components restricted to subsets of the microbial communities can be utilized to define the community structure. The concept of 'signatures' for subsets of the community based on the limited distribution of specific components has been validated by using antibiotics and cultural conditions to manipulate the community structure. The resulting changes agreed both morphologically and biochemically with the expected results (White *et al.* 1980). Other validation experiments involved isolation and analysis of specific organisms together with their subsequent detection in mixed culture experiments, utilization of specific inhibitors with the appropriate response, and the specific responses to changes in the local environment such as the light intensity. These validation experiments are summarized in a review (White 1983).

Phospholipids are found in the membranes of all cells. Under the conditions expected in natural communities the bacteria contain a relatively constant proportion of their biomass as phospholipids (White *et al.* 1979b). Phospholipids are not found in storage lipids and have a relatively rapid turnover in some sediments so the assay of these lipids gives a measure of the 'viable' cellular biomass (White *et al.* 1979a).

The ester-linked fatty acids recovered from the phospholipids (PLFA) are presently both the most sensitive and the most useful chemical measures of microbial biomass and community structure thus far developed (Bobbie and White 1980; Guckert et al. 1985). The specification of fatty acids that are ester-linked in the phospholipid fraction of the total lipid extract greatly increases the selectivity of this assay as most of the anthropogenic contaminants as well as the endogenous storage lipids are found in the neutral or glycolipids fractions of the lipids. By isolating the phospholipid fraction for fatty acid analysis it proved possible to show bacteria in the sludge of crude oil tanks. The specificity and sensitivity of this assay has been greatly increased by the determination of the configuration and position of double bonds in monoenoic fatty acids (Nichols et al. 1985; Edlund et al. 1985) and by the formation of electron capturing derivatives which after separation by capillary GLC can be detected after chemical ionization mass spectrometry as negative ions at femtomolar sensitivities (Odham et al. 1985). This makes possible the detection of specific bacteria in the range of hundreds of cells. Since microbial consortia from many environments such as marine sediments often yield 70 ester-linked fatty acids derived from the phospholipids, a single assay provides a large amount of information. Combining a second derivatization of the fatty acid methyl esters to provide information on the configuration and localization of the double bonds in mono-unsaturated components provides even deeper insight (Nichols et al. 1986). By utilizing fatty acid patterns of bacterial monocultures, Myron Sasser of the University of Delaware in collaboration with Hewlett Packard has been able to distinguish between over 8000 strains of bacteria (Sasser 1985). Thus analysis of the fatty acids can provide insight into the community structure of microbial consortia as well as an estimate of the biomass.

Despite the fact that the measurement of PLFA cannot provide an exact description of each species or physiologic type of microbes in a given environment, the method provides a quantitative description of the predominating microbiota in the particular environment sampled. With the techniques of statistical pattern recognition analysis it is possible to provide a quantitative estimate of the differences between samples with PLFA analysis. The sulfate-reducing bacteria are of particular importance in MFC (Pope et al. 1984). These organisms contain lipids which can be utilized to identify at least a portion of this class. Some contain a unique profile of branched saturated and mono-unsaturated PLFA (Edlund et al. 1985; Parkes and Taylor 1983; Taylor and Parkes 1983, Dowling et al. 1986) that allows differentiation between those utilizing lactate and those using acetate and higher fatty acids. Detailed analysis of the PLFA recovered from sulphate-reducing bacteria in this laboratory strongly suggests that the majority of sulphate-reducing bacteria found in marine sediments and in waters used in the secondary recovery of oil are the acetate-utilizing strains.

The effects of these 'obligate anaerobic' bacteria in essentially aerobic environments are frequently both spectacular and extensive (Puckorius, 1983).

How are these obligate anaerobes capable of operation in apparently aerobic environments? Hamilton (1985), among other workers, envisages a biofilm effect where sulphate-reducing bacteria only actually function when protected from the predominantly aerobic environment by voracious oxygen consumers and facultative anaerobic consortia. How then do the bacteria withstand the rigors of oxygen exposure in transit to the anaerobic site? Hardy and Hamilton (1981) demonstrated that five strains of the sulphate-reducing bacterium *Desulfovibrio vulgaris* were able to remain viable for over 72 hours of oxygen stress in aerated water due, at least in part, to the presence of the enzymes superoxide dismutase and catalase. Subsequently Cypionka *et al* (1985) showed that other genera of sulphate-reducers, including the acetate-oxidizing *Desulfobacter*, were also resistant to oxygen stress.

If indeed sulphate-reducing bacteria do exhibit oxygen tolerance in the environment, as well as in monocultures, then a constant injection of cells into the aerobic mainstream in order to colonize new substrata would provide the inoculum. However, in order to grow and by their metabolic activity facilitate corrosion, they must exist in an anaerobic microniche. Not only do the sulphate-reducers rely upon other bacteria to remove oxygen but also to supply carbon and energy sources.

In this study evidence is presented from preliminary experiments for increased corrosion rates associated with the presence of sulphate-reducing bacteria in coculture with facultatively aerobic fermenting bacteria in aerobic seawater medium. Data is also presented which demonstrates that the characteristic PLFA of *Desulfovibrio* type sulphate-reducing bacteria are found in thin biofilms formed on stainless steel surfaces exposed to rapidly flowing highly-aerated seawater. This shows the rapid recruitment of these organisms to various metal surfaces exposed to aerobic seawater.

EXPERIMENTAL

Experiments were set up to determine if enhanced corrosion of 304 stainless steel coupons could be detected in cocultures of *Vibrio natriegens* and the sulphate-reducing bacterium *Desulfobacter postgatei* strain 2ac9 when compared to cultures of *Vibrio natriegens* alone. In the first experiment the bacteria were cultured in half strength 2216 Difco marine broth in Erlenmeyer flasks. (It had previously been determined that

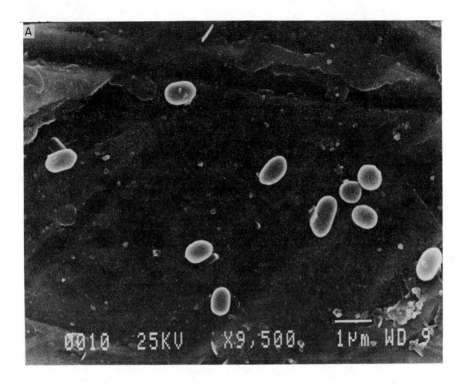

FIG 1(a)   Scanning electron micrographs of (a) bacterial cells attached to a steel coupon from batch cultured *V. natriegens*, (b) cells and glycocalyx attached from chemostat cultured *V. natriegens*, and (c) cells and glycocalyx from chemostat cocultured *V. natriegens* and *Desulfobacter postgatei*.

Desulfobacter sp. could grow satisfactorily in 2216 MB when the medium was reduced with sodium sulphide and provided with acetate as carbon and electron source). Duplicate flasks were set up for each treatment, each flask containing five coupons (four for corrosion analysis, and one for scanning electron microscopy, SEM). The different treatments were:

1. Vibrio inoculated alone with the flask sealed with cotton wool and incubated with rotory shaking (aerobic).
2. Vibrio and Desulfobacter inoculated with the flask sealed with cotton wool and incubated with rotory shaking (aerobic).
3. Vibrio and Desulfobacter inoculated with the flask sealed with a bung (limited oxygen).

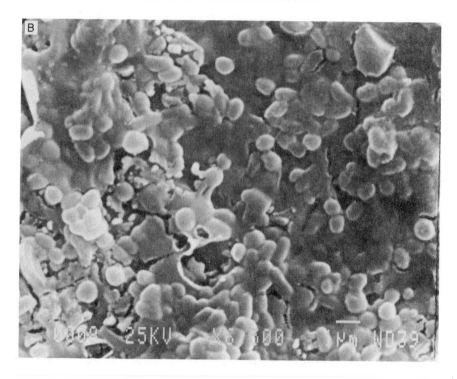

FIG 1(b).

The flasks were incubated for four weeks at room temperature on an orbital shaker. At the end of this period four coupons from each flask were analysed by an EG & G PARC model 350 A corrosion measurement console for Tafel constants and polarization resistance from which Icorr. values were calculated as described (Nivens *et al.* 1986). The remaining coupons were examined by SEM after critical point drying and shadow contrasting.

The treatments had the following average Icorr. values: 1) $6.5\,nA/cm^2$, 2) $10.4\,nA/cm^2$, 3) $23.0\,nA/cm^2$ (control with no bacteria was $4\,nA/cm^2$). The SEM micrographs showed that very few cells remained attached to the coupons after the latter were removed from the media (Figure 1). The Icorr. values appear to indicate an extremely small enhanced corrosion rate with the presence of a coculture with *Desulfobacter*.

The second experiment was designed to observe any differences in corrosion rates in organisms selected to form a biofilm on the stainless steel coupons. The organisms were selected in a continuous culture apparatus

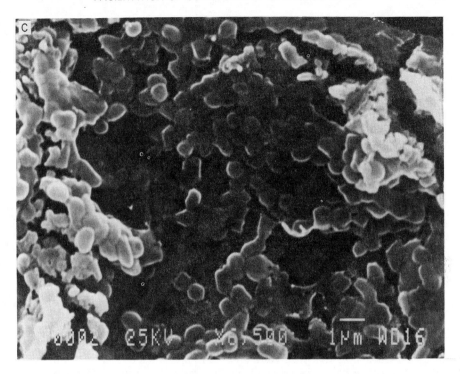

(Figure 2). The nutrient content of the media feed was decreased 10-fold and was supplied at a rate of 1.5 ml/min./ reaction vessel. The working volume was 800 ml with a dilution rate of 0.1/hr. The medium used was 1/10th strength 2216 Difco marine broth diluted in 'Forty fathoms' marine salts. The continuous culture system was run for one month after which the coupons were recovered and analysed as in the batch culture experiment.

SEM micrographs of the chemostat coupons showed that bacteria were more adherent under these conditions than those of the batch experiment, possibly correlating with the large quantities of glycocalyx produced (Figure 1). Corrosion values were also elevated: Coupons incubated with *V. natriegens* had a mean Icorr. value of 25.5 nA/cm². Coupons incubated with both *Vibrio* sp. and *Desulfobacter* sp. had a mean Icorr. value of 93.3 nA/cm². It still remains, however, to be shown that these differences in corrosion rates were solely due to the metabolic activity of sulphate-reducing bacteria. Unfortunately the SEM micrographs showed no

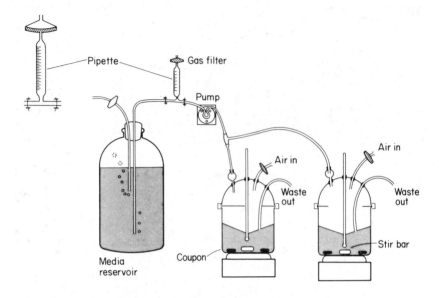

FIG 2   Continuous culture apparatus used to culture *Vibrio natriegens* and coculture *V. natriegens* and *Desulfobacter postgatei*. Both culture vessels contained 304 stainless steel coupons which were analysed after one month of the apparatus's operation to determine if the presence of the sulphate-reducing bacterium enhanced the corrosion rate.

difference in cell morphology in the biofilm. It may be unreasonable, however, to expect to observe a difference when the largest concentration of metabolically active sulphate-reducing bacterial cells are expected to be under the Vibrios in anaerobic microniches. Further experiments are in progress to study the mechanism of the enhancement of corrosion.

From the above two experiments it appears reasonable to expect recruitment of sulphate-reducing bacteria to metal surfaces in the aerobic, oligotrophic marine environment. At the IFREMER laboratories in Brest, France, several types of metal tubing were exposed to seawater over periods of 2, 5, 10, 15 and 30 days (Guezennec, 1985). Seawater was passed through the tubes at varying rates, from 0.1 to 1.5 m./sec. After exposure to seawater the lipids were extracted from the biofilm with a one phase chloroform/methanol/water mix (Bligh and Dyer, 1959), fractionated by silicic acid column (Gehron and White, 1983) and the phospholipids recovered. Mild akaline methanolysis released the PLFA as methyl esters. The PLFA methyl esters were characterized

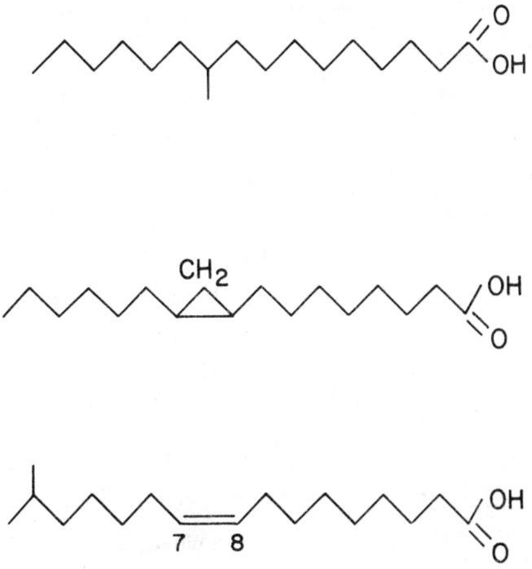

10 Me 16 : 0
(Δ^{10}-methyl hexadecanoic acid)

Cyclopropyl 17 : 0
(Δ^9, Δ^{10}-methylene hexadecanoic acid)

iso 17 : 1ω7cis
(Δ^{15}-methyl-Δ^9, Δ^{10}-hexadecanoic acid)

FIG 3   Fatty acid biomarkers used to detect members of the genera *Desulfobacter*
and *Desulfovibrio*.

and quantified by gas chromatography and gas chromatography-mass
spectrometry.

Several workers including Boon *et al.* (1978), Taylor and Parkes (1983),
Edlund *et al.* (1985), and Dowling *et al.* (1986) have proposed that the
fatty acids found in the *Desulfovibrio* spp. and *Desulfobacter* spp.
may be used as biomarkers in the environment. These biomarkers include
iso 17:1ω7cis (15 methyl-9, 10 hexadecanoic acid) and anteiso 17:1ω7cis
(14 methyl-9, 10 hexadecanoic acid) for members of the genus *Desulfovibrio*, and 10Me16:0 (10-methyl hexadecanoic) and cyclopropyl (cy) 17:0
(9, 10-methylene hexadecanoic acid) for members of the genus
Desulfobacter.

Examination of the biofilm fatty acid spectra obtained from aluminium, steel and titanium tubes generally showed the appearance of the
fatty acid iso 17:1ω7cis within a two-day period, and the fatty acids
10Me16:0/Cy 17:0 within a five-day period in the summer (April-May,
1985). Confirmation of the presence of sulphate-reducing bacteria was only
attempted from 30-day biofilms however. From these biofilms hydrogen-
and acetate-oxidizing sulphate-reducing bacteria were enriched which were
presumed to be members of the genera *Desulfovibrio* and *Desulfobacter*
respectively.

DISCUSSION

The preliminary experiments reported here suggest that a consortium of
the facultatively anaerobic fermenter *V. natriegens* and the obligate

anaerobe *D. postgatei* in consortia when in an attached biofilm can facilitate the corrosion of stainless steel exposed to aerobic seawater. With the use of the non-destructive, Fourier transforming infrared spectrometry (FT/IR) for the analysis of biofilms produced by *V. natriegens*, Nivens *et al.* (1986) was able to show that the accumulation of the polysaccharide exopolymer containing calcium hydroxide induced a reversible acceleration of corrosion of stainless steel exposed to seawater. The fact that the consortium increased the corrosion suggests that the activity of the sulphate-reducing bacteria may be involved. Preliminary experiments have shown that the Vibrio forms butanol with traces of acetate during anaerobic catabolism of glucose. The fact that the sulphate-reducer grew (evidenced by the decrease in sulphate in the batch culture) indicates that traces of acetate were utilized by the Desulfobacter. It also suggests that the addition of another fermenter which can form acetate from butanol should potentiate the growth of the sulphate-reducer and the corrosion. Since the analysis of 'signature' PLFA involves the isolation and identification of specific fatty acid methyl esters by GC/MS, it is possible to use 13-C mass-labelled precursors to follow the incorporation into the 'signature' biomarker lipids of the various component members of the consortia as has been done for muramic acid (Findlay *et al.*, 1983).

Not only can the details of interactions between components in microbial consortia be examined in detail, but the understanding of mechanisms of potentiation of corrosion can be explored. Common substrates for the sulphate-reducing bacteria include lactate, carbon dioxide, acetate, propionate, and molecular hydrogen. Removal, and in some cases generation, of these molecules by sulphate-reducers may contribute to their corrosiveness (Pope *et al.*, 1984). The by-product hydrogen sulphide may also accelerate corrosion. Some members of the genus *Desulfovibrio* contain hydrogenases. The classic mechanism by which sulphate-reducing bacteria potentiate corrosion is by cationic depolarization with the removal of hydrogen (von Wolzogen Kuhr and Van der Vlugt (1934). Iverson (1982) has suggested that labile iron phosphide compounds are the primary factors in the anaerobic corrosion process in monocultures of *Desulfovibrio desulfuricans*. The potentiation of corrosion by the *Desulfobacter postgatei* suggested by the preliminary data in this study make clear that at least with this sulphate-reducing bacteria cationic depolarization cannot be a factor as it contains no detectable hydrogenase. MFC may prove an excellent tool with which to study the interactions of microbial consortia in biofilms. It is clear that the creation of anaerobic microniches by the metabolic activities of consortia of bacteria in biofilms in aerobic systems (White, 1986) must be considered in the attempts to control corrosion of structures in the sea with its high sulphate concentration.

ACKNOWLEDGEMENTS

This work was supported by contracts N0014-82-C0404 and N0014-83-0056 from the Department of the Navy, Office of Naval Research.

LITERATURE CITED

Bligh, E. G. and Dyer, W. G. (1959). 'A rapid method of total lipid extraction and purification', *Canad. J. Biochem. Physiol.*, **37**: 911–917.

Bobbie, R. J. and White, D. C. (1980). 'Characterization of benthic microbiol community structure by high resolution gas chromatography of fatty acid methyl esters', *Appl. Environ. Microbiol.*, **39**: 1212–1222.

Boon, J. J., Liefkens, W., Rijpstra, W. I. C. G., Baas, M. and De Leeuw, J. W. (1978). 'Fatty acids of *Desulfobacter desulfuricans* as marker molecules in sedimentary environments', in *Environmental biogeochemistry and geomicrobiology. Volume 1: The aquatic environment* (Ed. W. E. Krumbein.), pp. 355–372, Ann Arbor Science Pub. Inc., Michigen.

Cypionka, H., Widdel, F. and Pfennig, N. (1985). 'Survival of sulphate-reducing bacteria after oxygen stress, and growth in sulphate free oxygen-sulphide gradients', *FEMS Microbiol. Ecol.* **31**: 39–45.

Dowling, N. J. E., Widdel, F. and White, D. C. (1986). 'Analysis of phospholipid ester-linked fatty acid biomarkers of acetate-oxidizing sulfate reducers and other sulfide forming bacteria', *J. Gen. Microbiol.*, **132**: in press.

Edlund, A., Nichols, P. D., Roffrey, R. and White, D. C. (1985). 'Extractible and lipopolysaccharide fatty acid and hydroxy acid profiles from *Desulfovibrio* species', *J. Lipid Res.*, **26**: 982–988.

Findlay, R. H. D., Moriarty, J. W. and White, D. C. (1983). 'Improved method of determining muramic acid from environmental samples', *Geomicrobiol. J.*, **3**: 135–150.

Gehron, M. J. and White, D. C. (1983). 'Sensitive assay of phospholipid glycerol in environmental samples', *J. Microbial. Meth.*, **1**: 23–32.

Guckert, J. B., Antworth, C. B., Nichols, P. D. and White, D. C. (1985). Phospholipid, ester-linked fatty acid profiles as reproducible assays for changes in prokaryotic community structure of estuarine sediments', *FEMS Microbial. Ecology*, **31**: 147–158.

Guezennec, J. (1986). 'La colonisation bacterienne des surfaces metalliques exposees en milieu marin. Utilisation des lipides Bacteriens', PhD. thesis: Docteur de L'Universite Paris VI.

Hamilton, W. A. (1985). 'Sulfate-reducing bacteria and anaerobic corrosion', *Ann. Revs. Micro.*, **39**: 195–217.

Hardy, J. A. and Hamilton, W. A. (1981). 'The oxygen tolerance of sulfate-reducing bacteria isolated from North sea waters', *Curr. Microbiol.*, **6**: 259–262.

Iverson, W. P. (1983). 'Anaerobic corrosion mechanisms', *Corrosion '83*. National association of corrosion engineers. Paper 243.

Nichols, P. D., Smith, G. A., Antworth, C. P., Hanson R. S. and White, D. C. (1985b). 'Phospholipid and lipopolysaccharide normal and hydroxy fatty acids as potential signatures for the methane-oxidizing bacteria', *FEMS Microbial Ecology*, **31**: 327–335.

Nichols, P. D., Guckert, J. G. and White, D. C. (1986). 'Determination of monounsaturated fatty acid double-bond position and geometry for microbial monocultures and complex consortia by capillary GC/MS of their dimethyl disulfide adducts', *J. Microbiol. Meth.*, **4**: in press.

Nivens, D. E., Nichols, P. D., Henson, J. M., Geesey, G. G. and White, D. C. (1986). 'Reversible acceleration of corrosion of stainless steel exposed to seawater induced by the extracellular secretions of the marine vibrio *V. natriegens*', *Corrosion*, **42**: 204–210.

Odham, G., Tunlid, A., Westerdahl, G., Larsson, L., Guckert, J. B. and White, D. C. (1985). 'Determination of microbial fatty acid profiles at femtomolar levels in human urine and the initial marine microfouling community by capillary gas chromatography-chemical ionization mass spectrometry with negative ion detection', *J. Microbiol. Meth.*, **3**: 331–344.

Parkes, R. J. and Taylor, (1983). 'The relationship between fatty acid distribution and bacterial respiratory types in contemporary marine sediments', *Estuarine Coastal Mar. Sci.*, **16**: 173–189.

Pope, D. H., Duquette, D. J., Johannes, A. H. and Wagner, P. C. (1984). 'Microbio logically influenced corrosion of industrial alloys', *Materials performance*, April: 14–18.

Puckorious, P. R. (1983). 'Massive utility condenser failure due to sulphide-producing bacteria', *Corrosion '83*, National association of corrosion engineers, Paper 248.

Sasser, M. (1985). 'Identification of bacteria by fatty acid composition', *Am. Soc. Microbiol. Meet.*, March 3–7, 1985.

Tatnall, R. E. (1981). 'Fundamentals of bacteria-induced corrosion', *Corrosion '81*, National association of corrosion engineers.

Taylor, J. and Parkes, R. J. (1983). 'The cellular fatty acids of the sulfate-reducing bacteria, *Desulfobacter sp., Desulfobulbus sp.*, and *Desulfovibrio desulfuricans*', *J. Gen. Microbiol.*, **129**: 3303–3309.

von Wolzogon Kuhr, C. A. H. and Van der Vlugt, L. S. (1934). *Water*, **18**: 147.

White, D. C. (1983). 'Analysis of microorganisms in terms of quantity and activity in natural environments', in *Microbes in their natural environments*, J. H. Slater, R. Whittenbury and J. W. T. Wimpenny (eds.), *Society for General Microbiology Symposium* **34**: 37–66.

White, D. C., Davis, W. M., Nickels, J. S., King, J. D. and Bobbie, R. J. (1979a). 'Determination of the sedimentary microbial biomass by extractible lipid phosphate', *Oecologia*, **40**: 51–62.

White, D. C., Bobbie, R. J., Herron, J. S., King, J. D. and Morrison, S. J. (1979b). 'Biochemical measurements of microbial mass and activity from environmental samples', in *Native Aquatic Bacteria: Enumeration, Activity and Ecology*, J. W. Costerton and R. R. Colwell (eds.)., ASTM STP 695, American Soc. for Testing and Materials, pp. 69–81.

White, D. C., Bobbie, R. J., Nickels, J. S., Fazio, S. D. and Davis, W. M. (1980). 'Nonselective biochemical methods for the determination of fungal mass and community structure in estuarine detrital microflora', *Botanica Marina*, **23**: 239–250.

White, D. C., (1986). 'Quantitative physical-chemical characterization of bacterial habitats', in *Bacteria in Nature*, Vol. 2, J. Poindexter and E. Leadbetter, eds., Plenum Publishing Corp., N.Y., in press.

Wolin, N. R. (1979). 'The rumen fermentation: a model for microbial interactions in anaerobic ecosystems', *Adv. Microbial Ecol.*, **3**: 49–77.

FACTORS AFFECTING THE DURABILITY OF REINFORCED CONCRETE UNDER SEMI-STAGNANT OFFSHORE CONDITIONS

T. D. B. Morgan and A. D. Steele

Shell Research Limited, Thornton Research Centre, P.O. Box 1, Chester, CH1 3SH, England

SUMMARY

The saline waters held in platform legs and storage cells of concrete structures offshore can undergo substantial chemical modification if stagnation occurs and microbial activity develops. This presentation is concerned with the detailed effects of such modification on ordinary Portland Cement concretes, and of the related consequences for steel, notably reinforcing steel.

The phases comprising typical engineering concretes exhibit different resistance characteristics towards the complex modified environment. For example, the tricalcium aluminate in concrete surfaces can be modified by sulphides, as well as by chlorides and sulphates. The tetracalcium aluminoferrite phase can also be susceptible to sulphide attack. The two silicate phases, usually regarded as fairly inert, can undergo some attack, particularly under more acidic conditions. The various reactions will be discussed in some detail, and will include references to synergistic effects.

Work on the factors affecting corrosion of reinforcing steel has shown that sulphate is just as effective as chloride in depassivating embedded reinforcing steel, but corrosion rates are, of course, limited by low oxygen availability. The effects of other likely constituents of leg and cell water will also be included.

Ways of detecting surface attack of concretes will be addressed, and the importance of initial concrete quality will be emphasized as a prime means of retaining excellent concrete integrity.

BACKGROUND

In a previous paper (Morgan *et al.* (1983)), authored jointly with our colleagues at Sittingbourne Research Centre, we described how microbial activity could be implicated in the modification of saline water environments, within storage cells for example, and considered how the ensuing effects of these environmental modifications on concrete durability might best be monitored and studied.

The present paper deals largely with the interaction of these chemically-modified environments with concrete and steel. In particular, an attempt has been made to isolate variables in these complex processes, thereby to establish the nature of fundamental processes that might affect concrete performance. Accordingly, the effects of various chemical variables both on steel and concrete performance have been evaluated.

MATERIALS OF INTEREST

In the context of engineering structures, it is the concretes based on ordinary Portland cement (OPC) that have been used most widely. Often OPC concretes are supplemented by quantities either of pulverized fuel ash (PFA) or blast furnace slag (BFS) in the mix design, not only in the interests of economy but also to (i) promote greater resistance to detrimental reactions between alkali and some aggregates (by mopping up excessive quantities of free lime), (ii) dilute heat liberation in the early stages of curing, and (iii) decrease porosity and therefore promote impermeability.

Whatever the detailed mix, these concrete types consist essentially of four phases, viz. tricalcium silicate, dicalcium silicate, tricalcium aluminate and tetracalcium aluminoferrite (Lea 1970)). These are often designated C_3S, C_2S, C_3A, C_4AF respectively. There are other products of the cement hydration and subsequent reactions that need to be considered in relation to durability, in particular free lime and those entities produced by the chemical action of aggressive species on each of the four phases above.

Reinforcement of steel typically used in concretes is basically a carbon steel, with manganese (at 0.75%w) as the only element except iron having a concentration exceeding 0.20%w. Specifications are set against mechanical properties criteria.

The experimental programme has been based on OPC concretes and reinforcing steels conforming to the above.

ENVIRONMENTS

The study has been extended well beyond a narrow window of conditions that would represent expected operational circumstances. The approach has been to investigate the influence of a full range of likely contaminants, sometimes in over-severe conditions so as to accelerate tests and establish trends quickly.

Since there is a need to consider the effects both of formation waters and seawater, depending on how the storage facilities are operated, it has been important to recognize firstly how differences in cation distribution can bring about quite different sulphate solubilities for each water type, and then also to recognize possible complications arising from mixing seawater and formation water.

Apart from the usual constituents of saline waters, the effects of hydrogen sulphide, carbon dioxide, acetic acid and sulphuric acid have been of central interest, reflecting metabolic processes. Similarly, the effect of pH has been studied, as has the effect of another important cathodic variable, viz. oxygen activity.

CONCRETE/ENVIRONMENT INTERACTIONS

Whilst the changes observed in physical and mechanical properties of concretes are of great relevance in traditional engineering terms, there is a wealth of telling information to be derived from chemical evaluation techniques. The range of non-destructive methods available was covered in a previous paper (Morgan et al (1983)).

Analytical methods used (and occasionally developed) to follow the progress of concrete exposure have included (i) X-ray diffraction of surface products, (ii) chloride permeation as a guide to permeation rates, (iii) X-ray mapping of element distribution and (iv) alkalinity profiling by indicator. Two chemical evaluations that have proved particularly useful have been (a) free lime depletion and (b) sulphoaluminate formation.

Considering firstly the effects of expected saline water constituents, the sulphate anion is of major relevance. This lies in the fact that the product of the reaction of sulphate with C_3A, i.e. sulphoaluminate, occupies a greater volume than its precursor; consequently, a notable physical change follows sulphaluminate formation (Chatterji et al. (1967)) and the well-known spalling effect results. Chloride attack of C_3A is not kinetically disfavoured by comparison with sulphate attack; it is merely the case that comparable volume changes do not occur on reaction, and therefore the concrete can accommodate chloride attack more readily.

If the environmental pH is allowed to drop below neutrality, then the concrete may start to use the alkalinity on which its stability depends. Obviously, whether such a process is significant depends upon the pH level attained, the time for which it is maintained, and the intrinsic reactivity of the concrete under consideration.

The nature of the acidity is also important. Accordingly, acetic acid, which forms a highly soluble calcium salt, can bring about significant attack of certain mortars/concretes by increasing porosity as reacted free lime dissolves in the acid containing medium. Sulphuric acid at an equivalent pH can be less aggressive because, for example, the sulphate product of free lime depletion is more insoluble, hence restricting porosity and thereby the progress of attack.

PLATE 1   Showing acid attack of mortar sample (top), unexposed mortar sample (bottom).

Acidic attack follows quite a different route from sulphate attack, creating porosity and also damaging all the concrete phases when free lime depletion is effectively complete. In this second process, C_3A probably remains the most vulnerable phase, but there is evidence for the breakdown of polysilicate chains in C_2S (Rodger (1985)). The characteristics of the two attack modes are evident from Plates 1 and 2.

In situations where formation waters containing a few hundred ppm of acetates are acidified by mineral acidity, it is important to recognize the potentially increased potency of the aqueous medium since acetic acid may well be the aggressive agent of note.

The effect of dissolved CO_2 is really no more than a pH effect except that surface carbonation can result. H_2S in isolation has been found to be relatively benign towards concrete. However, in the latter regard, it should be noted that attack of C_3A and C_4AF phases is possible (see later), C_4AF attack being signalled by a colour change to black, reflecting iron sulphide formation. Furthermore it should be noted that consecutive exposure to H_2S and air might provide a route to sulphur oxyacid attack of concrete, e.g. to sulphoaluminate formation.

PLATE 2   Showing sulphate attack of mortar sample (top), unexposed mortar sample (bottom).

As the combined effects of mixtures of contaminants have been evaluated, various important synergistic effects have emerged. The relative aggressiveness of (i) H_2S and acetic acid and (ii) sulphate and acetic acid are particularly noteworthy; but the influence of pH is quite important here, too. Consistent with the experimental findings is the idea that increased deterioration of surfaces will arise if different mechanisms of attack operate concurrently. Thus, in a situation where moderate acidity (say pH 5.0) is achieved in a water whose acetate content is 500 ppm, there is an initial neutralization of surface free lime and a mechanistic assistance (by acetic acid attack) for the increase of concrete porosity. In this situation, sulphate can gain far improved access to unreacted tricalcium aluminate, form sulphoaluminate, and thereby exert a synergistic rather than additive influence.

Similar considerations apply to the observed synergism between H_2S and acetic acid at moderate pH, with or without oxygen being involved. The combination of lowered pH and acetate attack allows more effective attack by the H_2S, probably the C_3A and C_4AF. It has become clear that the sulphiding reaction can, itself, be damaging to surfaces of concrete in these circumstances, even without the involvement of oxygen. Should oxygen be introduced, a lightening of the concrete colour is evident as sulphide is oxidized; the analysis shows that sulphoaluminate can be formed by such a route (see the related examples in Table 1), and there is also substantial formation of elemental sulphur. Visual assessment indicates attack typically associated with sulphoaluminate formation.

Table 1   Effect of H_2S on sulphoaluminate formation on concrete surfaces

		[Sulphoaluminate] %w
Simulated seawater pH 5.0; adjusted with sulphuric acid	Concrete 1	0.96
	Concrete 2	0.89
As above, with added H_2S to 200 ppm. Oxidation then allowed.	Concrete 1	1.13
	Concrete 2	1.16

N.B.   Sulphoaluminate measurement is made on sample drilled from first 5 mm of exposed block. Exposure time = 12 months.

Where oxygen is not involved, the mechanism whereby concrete might deteriorate has yet to be fully characterized. However, it is clear that a complex sulphide species may be formed, and aluminosulphides and complex iron sulphides are possibilities.

Interestingly, and as important evidence for the mechanisms suggested, equivalent experiments at pH 6.0 show little tendency for attack of concrete, even though there are indications of some free lime depletion at surfaces, and also of some sulphoaluminate formation. This observation is important in relation to the possible need for identification of control options (see later).

STEEL/ENVIRONMENT INTERACTIONS

Considerations of the susceptibility of embedded rebar steel to attack by permeating saline fluids have been the subject of various reviews. Pitting attack of steel, usually identified with the depassivation of the steel surface by chlorides, can be promoted if a suitable cathodic driving force is present. In practice, this means that a sufficient oxygen flux is necessary.

Using electrochemical techniques, notably, d.c. polarization studies and a.c. impedance where appropriate, depassivation of steel has been examined both in saturated lime solutions — simulating pore waters — and in the solid state.

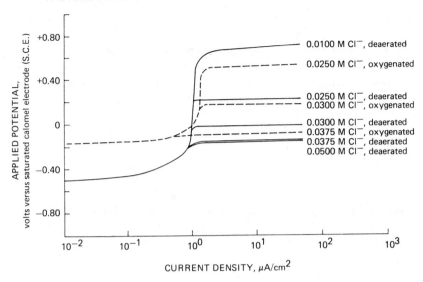

FIG 1   Potentiodynamic anodic polarization of rebar steel in saturated (0.0250 M) Ca(OH)$_2$ solution containing various [Cl$^-$], oxygenated and deaerated. Sweep rate = 0.1 mV/s.

Data given in Figure 1 demonstrate that for a pH of 12.5, the [Cl$^-$]/[OH$^-$] ratio required to depassivate the steel is 0.75 M corresponding to a [Cl$^-$] concentration of 0.0375 M. This compares favourably with the figure of 0.61 suggested by Hausmann (1967). However, by far a more significant point is that sulphate can be similarly aggressive towards passive steel in the absence of chloride; results to date suggest the critical [SO$_4^=$]/[OH$^-$] ratio is 0.42 at pH 11.5.

This type of observation rationalizes the quite significant differences in the behaviour of acetic acid and sulphuric acid when each has been used separately to investigate depassivation tendencies as pH is reduced. Addition of acetic acid to saturated lime solutions can be continued to a pH below 10 before the polarization curves indicate any clear loss of passive characteristics. Conducting the same experiment with sulphuric acid, expected in practice if oxidative reactions prosper, allows only a minor reduction of pH, to 11.5, before the loss of passive characteristics is readily achieved. Immersion tests confirm these trends completely.

The known problem areas for steels in H$_2$S or CO$_2$ service do not signify to any extent in the present context. The low concentrations of these species coupled with the absence of significantly acidic conditions are the prime reasons. Some blistering of reinforcing steel was observed at lower pH values, but cracking phenomena have been absent.

As the more complex mixtures have been looked at, some interesting effects have again emerged. For instance, the presence of H_2S even at 50 ppm serves to assist the film-breaking characteristics of the chloride and sulphate species present in saline waters. Presumably, this is a consequence of sulphides being incorporated into the passive oxide/hydroxide film, thereby lowering its protecting capacity against chloride or sulphate attack.

It should also be recognized that, in the equivalent situation to that considered previously for concrete, i.e. moderate pH set by sulphuric acid added to a seawater/formation water mixture, any steel contacting such a solution would be prone to corrosion providing cathodic oxygen reduction or hydrogen evolution was maintained. However, to assume that this would happen in reality is to ignore the importance of concrete cover in moderating the environment, and (in a more specialized case) to underrate the self-healing characteristics of concretes where fine cracks are concerned.

CONSEQUENCES FOR REINFORCED CONCRETE DURABILITY

Of the various effects considered previously, it becomes clear that pH reduction and sulphate attack present potential problems to both concretes and steels. The consequences of surface attack by these entities, when combined, are increased porosity, free lime depletion, sulphoaluminate formation and spalling (for concretes), and lowered resistance to film breakdown (for steel). If the pH at the steel became low enough (reflecting a serious situation) then hydrogen evolution could supplement oxygen reduction as the cathodic half-cell reaction in a corrosion process.

The sharply different sulphate concentration present in seawater, compared with formation water, makes seawater potentially far more aggressive in terms of sulphate attack, at a given pH. However if formation waters contain significant quantities of acetate and acetate-homologue salts, there is a possibility that increased aggressiveness arising from acetic acid type attack counterbalances the loss of sulphate anions. In mixtures of such formation water and seawater, synergistic effects are probable, as previously described, even though some sulphate precipitation is to be expected.

Chlorides are far less a problem to concretes than are sulphates. Nevertheless, chlorides will depassivate steel if low-sulphate waters reach the steel-concrete interface. Again, it is the rate of the cathodic process that controls corrosion.

While CO_2 does not carry any marked influence, the overall effects of H_2S can be significant. Whether in relation to steel or concrete integrity, H_2S can be involved in significant ways in combination with acetic acid, chlorides, sulphates and even oxygen. However, the clear evidence that there is a sensitivity of rates of attack of concrete and steel to pH immediately points the way to any control measures, should they become necessary.

The optimism, that control measures may well not be needed, derives from one major factor protecting the integrity of concretes viz. low permeability. The practical engineering materials are chosen so that quality is directly tied up with high impermeability. It is this feature that has restricted damage to the surface layers, even over several years, i.e. to well within 10% of the total concrete cover. Such information has been collected in laboratory studies and in offshore monitoring exercises using techniques developed for the experimental study. Free lime profiles have been particularly useful in this area. Thus, by ensuring diffusion is restricted to low levels, concrete durability can be maximized because liquids cannot penetrate to any extent. The same line of defence would serve to limit oxygen diffusion, avoiding corrosion.

The work has therefore been important from a number of view points. The effects of the various compositional variables on materials performance have been evaluated. By knowing the important environmental variables, information is available that could be used to control the environment's aggressiveness to an acceptable level. Perhaps most importantly, the relevance of engineering practice to durability considerations has been established, providing every indication that properly chosen engineering criteria can result in full structural integrity despite the demands made by offshore oil and gas service.

ACKNOWLEDGEMENT

The authors are pleased to acknowledge the contributions made by colleagues at Thornton Research Centre, particularly by Mrs S. J. Barritt who has conducted much of the experimental work.

REFERENCES

Chatterji, S. and Jeffery, J. W. (1967). *Magazine of Concrete Research*, **6**, 185.
Housemann, D. A. (1967). 'Steel corrosion in concrete', *Materials Protection*, **6**, 19–23.
Lea, F. M. (1970). *The chemistry of cement and concrete*, Arnold, London.
Morgan, T. D. B., Steele, A. D., Gilbert, P. D. and Herbert, B. N. (1983), 'Concrete durability in acidic stagnant water', *Microbial corrosion conference*, Metals Society, NPL London, March 8–10.
Rodger, S. A. (1985), D. Phil. thesis, University of Oxford.

BIODEGRADATION OF CRUDE OILS IN RESERVOIRS

J. Connan

Elf Aquitaine

INTRODUCTION

Post-accumulation processes may drastically modify the original properties of crude oil pooled in a reservoir. Petroleum may still continue to change due to variations in temperature and pressure (subsidence of the reservoir), through the injection of gaseous fluids (hydrocarbons and CO_2) or through contact with meteoric waters carrying bacteria leading to water washing and biodegradation (Connan, 1984).

The biodegradation of petroleum or its alteration by bacteria is a widespread phenomenon in nature. The importance of the amount of petroleum affected by biodegradation needs to be appreciated. Demaison (1977) estimated that the seven largest accumulations of biodegraded petroleum including the Athabasca tar sands and the Orinoco tar belt in Venezuela, are together equivalent to the total volume of petroleum to be found in the 264 largest conventional crude oil fields (API gravity $> 20°$). The Orinoco tar belt alone, thought to be one of the largest extra heavy petroleum reserves (API gravity $< 10°$), contains between 700×10^9 and 1000×10^9 barrels of petroleum (1000 barrels = 159 cubic metres).

The biodegradation of petroleum increases its density (or reduces the API gravity). A petroleum is heavy if its density lies between $10°$ and $20°$ API. When the API gravity is lower than $10°$, the petroleum is classified as being extra heavy. Heavy oils, entailing complicated problems for producers, result from a combination of factors: evaporation of light ends (gas and gasoline fractions), biodegradation of saturated hydrocarbons (or alkanes) and aromatics including sulphur-bearing structures (benzo- and dibenzothiophenes, etc) and the simultaneous enrichment of polar compounds (resins and asphaltenes).

BACTERIAL ACTION ON CRUDE OIL

The biodegradation of hydrocarbons is an oxidative process. Saturated and aromatic hydrocarbons are transformed by several metabolic steps into oxygenated derivatives (fatty acids, alcohols, ketones, phenols, etc) which

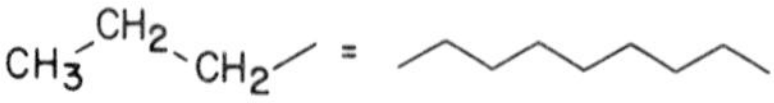

CH₃—CH₂—CH₂— =   (structure)

1. n - alkane

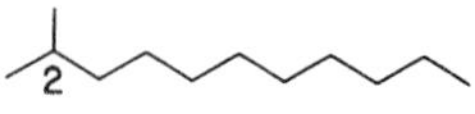

2. iso - alkane
(2 - methyl - alkane)

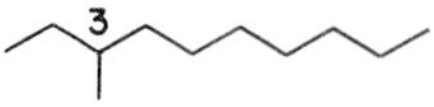

3. anteiso - alkane
(3 - methyl - alkane)

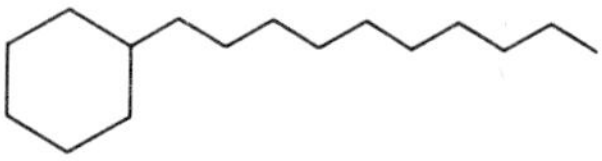

4. alkylcyclohexyl - alkane

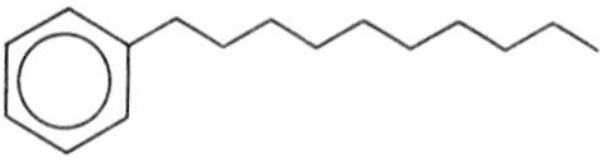

5. alkyl - benzene

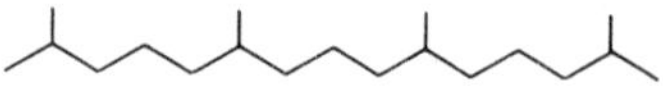

6. pristane (isoprenoid)

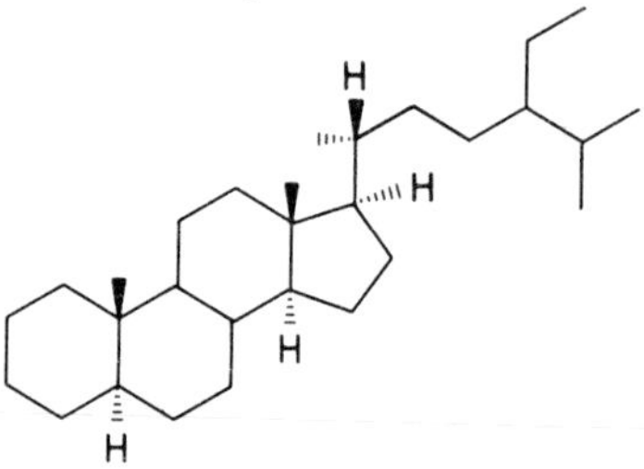

7. sterane (24 - ethylcholestane)

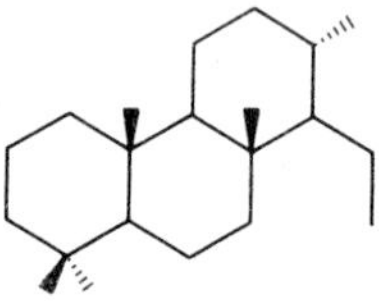

8. tricyclic terpane

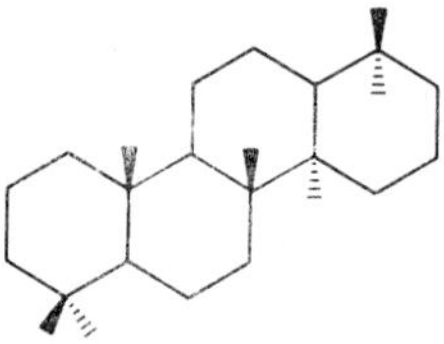

9. tetracyclic terpane

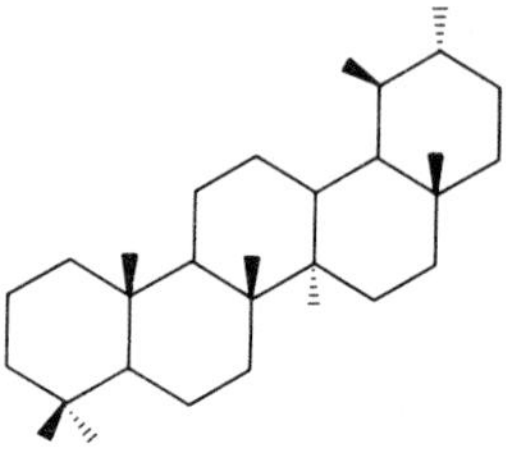

10. pentacyclic terpane
(lupane - 1)

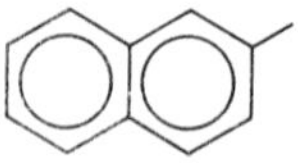

11. monomethylnaphthalene

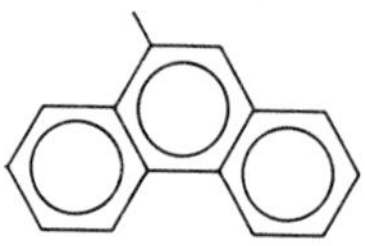

12. methyl - 9 phenanthrene

FIG 1   Chemical structures of molecules quoted in the paper. Example:
Alkylbenzenes (5).

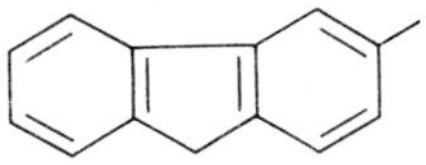

13. methyl-dibenzothiophene

14. cis-decalin

15. adipic acid

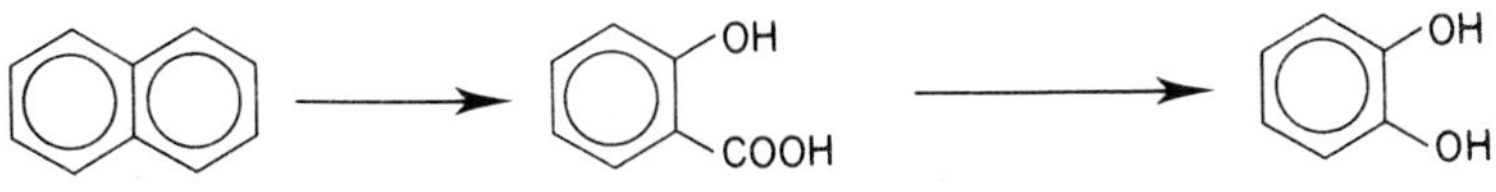

16. naphthalene

17. salicylic acid

18. pyrocatechol

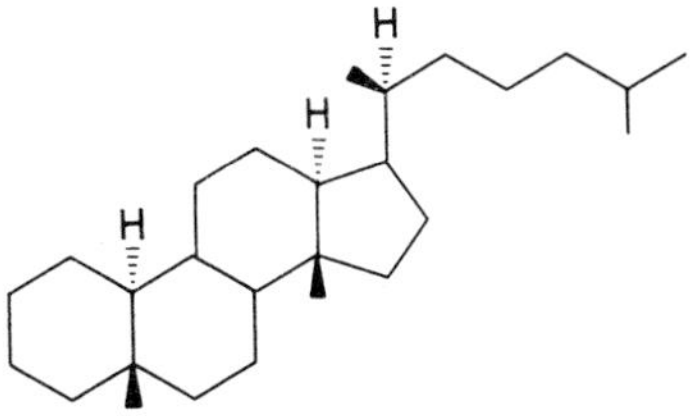

19. sterane (rearranged)

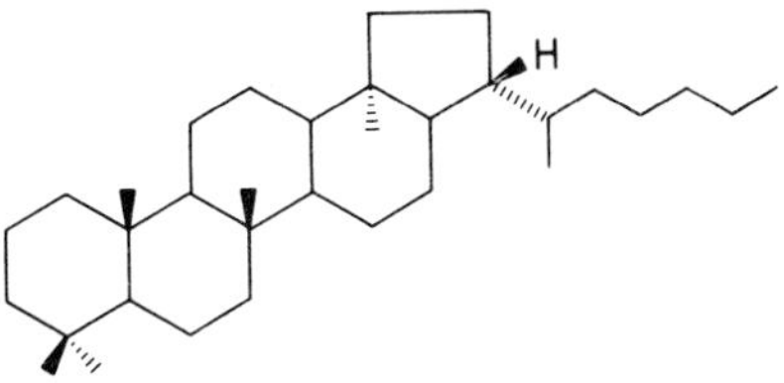

20. hopane (C_{35})

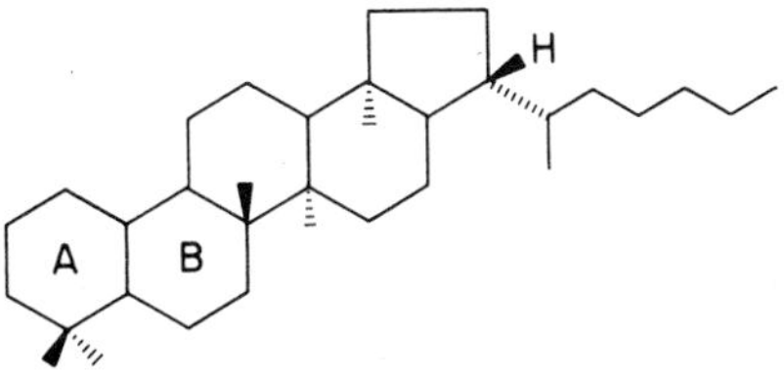

21. demethylated hopane (C_{34})

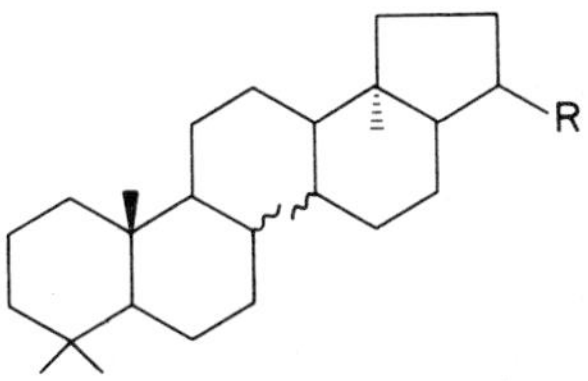

22. 8-14 secohopane

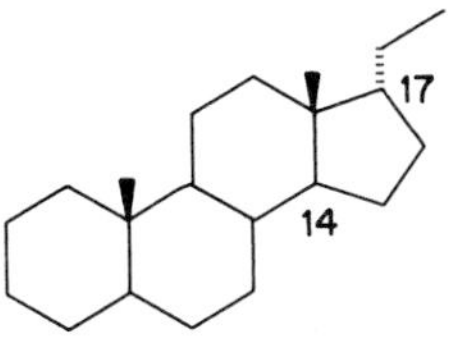

23. C_{21} sterane
$14\beta(H), 17\beta(H)$ – pregnane

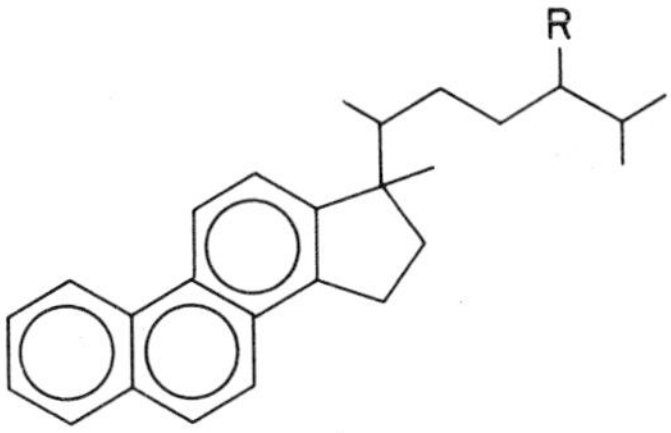

24. triaromatic steroid

FIG 1 *(Continued)*.

are then further broken down by α- and β-oxidation, ring-cleavage, etc. Biodegradation of n-alkanes, for instance, starts with an oxidation in the C_{-1} position giving an alkan-1-ol which is then converted to an alkan-1-oic acid. The reaction chain continues with α-β-oxidation or an oxidative decarboxylation.

Mechanisms involved in the biodegradation of hydrocarbons entail the formation of increasingly smaller molecules, ultimately to CO_2 and H_2O (in addition to biomass).

Many different micro-organisms are capable of attacking crude oil fractions but certain bacterial genera (*Pseudomonas, Corynebacterium* etc) are predominant in natural oil-degrading ecosystems.

The biodegradation of the lighter fractions of petroleum generates heavy petroleum residues, for instance, 'Coal tars' or the tar balls found on beaches.

The formation of these petroleum residues by weathering involves biodegradation of saturated and aromatic hydrocarbons with concentration of polar fractions (resins and asphaltenes). To these basic biological mechanisms can be added environmental factors, linked to marine and atmospheric conditions, eg evaporation of light fractions, photo-oxidation, chemical oxidation and water washing.

Under reservoir conditions, biological degradation of crude oil can only take place under specific circumstances:

- the presence of flowing water (essentially meteoric water)
- oil/water contact, since the bacteria only thrive in the aqueous phase
- oxygen
- a supply of nutrients (nitrate and phosphate)
- temperatures suited to the growth and activity of the bacteria.

The effects of biodegradation are detected in the hydrocarbon phase for petroleum pooled in reservoirs at temperatures between 20°C and 60–75°C.

THE VARIOUS STAGES IN THE BIODEGRADATION OF PETROLEUM: A SEQUENCE OF DEGRADATION OF INCREASINGLY COMPLEX STRUCTURES

In the first stages of the biodegradation of petroleum, bacteria selectively attack long chain structures (see Figure 1). n-Alkanes (*1*) are first to be degraded followed by iso- (2-methyl) and anteiso- (3-methyl) alkanes, alkylcyclohexyl- and alkylcyclopentylalkanes (*4*) and alkylbenzeness, in fact all the molecular structures, both saturated and aromatic, possessing a long alkyl chain.

When these molecules have disappeared, biodegradation of the alkanes continues with the removal of more branched structures (several methyl groups on the straight chain), the best known components being the isoprenoids (pristane *6*, phytane, etc).

Although linear and branched structures are easily assimilated, the same is not the case for cyclo-alkanes. Although ultimately degradable, these are comparatively poor substrates for micro-organisms. However, cis-decalin (*14*), for instance, is degraded by a *Flavobacterium* to adipic (*15*) and pimelic acids.

Steranes (tetracyclic *7*) and especially terpanes (tri- *8*, tetra- *9* and pentacyclic *10*) are extremely bioresistant. These hydrocarbons often survive in severely degraded petroleum where they may act as correlation markers, linking these oils to their unaltered counterparts. This does not, however, mean that they are not eventually degraded, since, like *n*-alkanes, they may also be destroyed, particularly in outcropping reservoirs.

Biodegradation also affects aromatic hydrocarbons. Alkylbenzenes (*5*) have already been quoted. Naphthalene (*16*) breakdown gives rise to monocyclic structures, eg salicylic acid (*17*), well known for its derivative aspirin, and catechol (*18*). More complex structures, namely alkylnaphthalenes (*11*), phenanthrenes (*12*) and sulphur-bearing aromatics (alkylbenzo- and alkyldibenzothiophenes *13*) are removed sequentially from crude oil.

BIODEGRADATION OF POLYCYCLIC ALKANES BY BACTERIA

Although the biodegradation of numerous alkanes (*n*-, iso-, anteiso-alkanes, isoprenoids, cyclohexyl-alkanes, etc) has been accepted for many years, evidence for the biodegradation of polycyclic alkanes was only put forward in the late 1970s. Seifert and Moldowan (1979) claimed that steranes and terpanes were degradable and proposed a biodegradability sequence among these particularly resistant structures. This biodegradation, initially deduced from observations in petroleum reservoirs, has been demonstrated under laboratory conditions by Goodwin *et al.* (1981).

In the molecular assemblage composed of tri-, tetra-, and pentacyclic alkanes, regular steranes (*7*) are more easily destroyed than are rearranged steranes or diasteranes (*19*). By focusing on regular steranes only, it can be ascertained that C_{27} steranes disappear before C_{29} steranes.

When the metabolism of steranes is initiated, pentacyclic alkanes of the hopane family (*20*) are the first to be degraded. This biodegradation preferentially depletes hopanes with a high molecular weight (C_{30}–C_{35}). Complete consumption of C_{29} and C_{30} hopanes occurs later when the diasteranes (*19*) disappear. It is frequently observed that the uptake of the hopane family (C_{29}–C_{35}) by bacteria leads to the formation of another chemically related homologous series: ring A/B dimethylated hopanes (21). This feature suggests that one of the biodegradation pathways for hopane structures is the removal of the C-10 methyl group in the A/B ring of the hopane (21).

The occurrence of a demethylated hopane series as degradation products of hopanes is not a general feature. Extremely biodegraded oils, as found in the Val de Travers asphalt, Switzerland, are devoid of these series. Have they themselves been eliminated during a more severe stage of biodegradation? The problem has still not been totally resolved. The residual alkanes do not show any traces of known classical structures: no steranes, hopanes, nor demethylated hopanes. Nevertheless, some compounds are still detectable, belonging to the 8-14 secohopane family (22). These compounds are concentrated progressively in the residual alkane fraction of severely biodegraded petroleum. From our present knowledge, it appears that these structures are particularly resistant to bacterial attack. They are still detectable when the tricyclic terpanes (8) and the $C_{21}-C_{22}$ steranes (23), considered to be the most resistant structures in crude oil, have been metabolized.

BIODEGRADATION STUDIES IN VITRO

To obtain a better understanding of the available observations on petroleum biodegraded under geological conditions, organic geochemists and microbiologists can attempt to reproduce some of these observations in laboratory experimentation (Restle, 1983). For this, in vitro simulations are made using either mixed cultures, enriched in bacteria living in soils polluted by hydrocarbons (refineries, filling sites, etc) or inocula of pure species of known hydrocarbon-degrading bacteria (Pseudomonas, Flavobacterium Achromobacter, etc). In the case of pure species of bacteria such as Pseudomonas oleovorans, micro-organisms are precultured for 24 hours at 25°C in a peptone-based medium.

In experiments using pure species (such as Pseudomonas oleovorans), petroleum fractions are incubated with non-growing cultures. Stationary phase conditions are obtained by a massive inoculum of precultured bacteria. The cultures are maintained at 25°C and are constantly shaken throughout the experiments (between 5 and 90 days). Mixed cultures, isolated from an oil-polluted soil may also be used in a similar way.

In vitro biodegradation with Pseudomonas oleovorans under non-growth conditions provides a satisfactory simplified model which allows reproduction of the first stages of hydrocarbon biodegradation in the laboratory. It has been observed that this species can degrade n-, iso-, anteiso- and cyclohexyl-alkanes (1, 2, 3, 4), C_9-C_{20} isoprenoids (6) and alkyl-benzenes (5).

Nevertheless, *in vitro* experiments are still an oversimplified system when compared with the wide experimental field offered in nature. Despite an experimental period of three months, the *Pseudomonas* species do not succeed in biodegrading polycyclic alkanes (steranes *6, 19, 23*; terpanes *8, 9, 10, 20* and aromatics (steroids *24*; methyl-phenanthrenes *12*; benzothiophenes).

Biodegradation of saturated polycyclic alkanes, difficult to obtain under laboratory conditions, was reported in 1981 by Goodwin *et al.* This successful experiment was fairly fortuitous. After one year, only two flasks out of a total of ten gave positive results. These results, although unexplained, demonstrated that the biodegradation of polycyclic alkanes can be reproduced in the laboratory. Since that report, further success has been achieved *in vitro*; preferential biodegradation of aromatics and intense biodegradation of polyaromatics including benzothiophenes have been established using microflora from present day marine sediments (Connan & Restle, 1984).

In conclusion, a firm understanding of the extreme biodegradation observed in petroleum reservoirs is expected to remain a far off target for some time to come. However, for the organic geochemist experimenting with hydrocarbon-utilizing bacteria, encouraging results which appear year after year bring this far horizon ever closer.

ACKNOWLEDGEMENT

The author is indebted to Dr. J. L. Shennan for her help in preparation of the final manuscript.

REFERENCES

Connan, J. (1984). 'Biodegradation of crude oils in reservoirs', in Brooks, J. and Welte, D. H. (eds): *Advances in Petroleum Geochemistry*, 299–335. Academic Press, London.

Connan, J. and Restle, A. (1984). 'La Biodégradation des hydrocarbures dans les resérvoirs', *Bull. Centres Rech. Explor.*, Prod Elf-Aquitaine, **8** (2) 291–302.

Demaison, G. J. (1977). 'Tar sands and supergiant oil fields', *Bull. Amer. Assoc. Petroleum Geol.*, **61**, 11, 1950–1961.

Goodwin, N. S., Park, P. J. and Rawlinson, A. P. (1983). 'Crude oil biodegradation under simulated and natural conditions', in Bjory, M. *et al.* (eds): *Advances in Organic Geochemistry 1981*, 650–658. Wiley, New York.

Restle, A. (1983). 'Etude de nouveaux marqueurs biologiques dans les pétroles biodégradés: cas naturels et simulations in vitro', Thèse Doctorat d'Etat. Univ. Louis-Pasteur, Strasbourg.

Seifert, W. K. and Moldowan, J. M. (1979). 'The effect of biodegradation on steranes and terpanes in crude oils', *Geochim. Cosmochim. Acta*, **43**, 111–126.

MICROBIAL DETERIORATION OF CORROSION PROTECTIVE COATINGS

Maria Stranger-Johannessen

Center for Industrial Research, Oslo, Norway

INTRODUCTION

Organic coatings are increasingly used to protect metal structures and equipment against corrosion. The coatings may be of quite different nature, based on oleoresins, chlorinated rubber, coal tar, polyurethane, polyamide, vinyl, epoxy resins and other high polymeric compounds.

To obtain a good performance of the coatings it is important that application rules are strictly adhered to. The metal surface must be clean and pinholes or 'holidays' must be avoided. Even if the coatings then have been thoroughly inspected and approved unexpected failures such as blistering and debonding are often experienced after a comparatively short time of service.

One of many examples is the failure of a seawater-carrying pipe system at a Norwegian metal working plant. The system consisted of 800 mm diameter steel pipes, internally coated with primer and three layers of polyamide. Heavy corrosion and water leaking through the pipe walls were observed 18 months after installation. Inspection of the pipes' interiors showed severe deterioration of the coating with blisters and areas where the coating had flaked off. In spite of comprehensive investigations, this early deterioration could not be explained.

It is not unusual that such failures cannot be explained. Montle, 1971, pinpoints the problem by stating: 'One particular type of blistering is not reproducible. Also, this phenomenon is puzzling in that it does not occur all of the time even with the same primer and the same topcoat.' and 'Nobody has a complete understanding of why bubbling of topcoats occurs and exactly how to completely eliminate topcoat bubbling'.

According to our investigations (Stranger-Johannessen, 1980, 1984, 1985), blistering and debonding of corrosion protective coatings can be caused by micro-organisms.

DETERIORATION OF COATINGS BY BACTERIA

The epoxy resin coating of a ship's tank for molasses, fatty oils and other fluid cargoes deteriorated and flaked off after nine months of service. The damage was particularly pronounced on horizontal areas where water remained after cleaning. The deterioration occurred again, and repeatedly, after sandblasting and recoating.

By microbiological investigation of the coating flakes, several types of bacteria could be isolated. Steel coupons with new coating were tested by incubation in suspensions of the single cultures. Two types of bacteria, both sporeforming *Bacillus*[sp.], caused brittleness and debonding of the coating film, while the remaining cultures did not cause any change.

The results were confirmed when properly coated steel rods were incubated in a mixed suspension of the two *Bacillus sp.* Single blisters appeared within one week, and the coating became densely blistered and debonded on longer incubation. No deterioriation occurred on the coating when it was incubated in a sterile medium.

This investigation shows that certain bacteria may be the cause of blistering and debonding of the corrosion protective coatings.

DETERIORIATION OF COATINGS BY FUNGI

The holds of a ship, carrying cereals, wood and other dry cargoes, became severely corroded within some months. Heavy pitting and reduced thickness of the steel plate were observed, and the steel surface was covered with corrosion products.

A microbiological investigation showed that the corrosion products were populated with viable fungi. These fungi caused corrosion of new steel coupons in laboratory tests, in the presence of a certain amount of nutrients.

Originally the holds' inner surfaces had been coated with a chlorinated rubber paint for corrosion protection. Laboratory tests with a corresponding paint on steel coupons showed that this paint provided just the nutrients necessary for the fungal growth. On incubation the fungi grew well on the coating surface and caused blistering and debonding within 10 days. Subsequently the paint flaked off and the fungi proceeded to grow on the steel surface. Pitting and other corrosion were found when the fungal growth was removed from the surface.

According to this investigation fungi may also cause the deterioration of coatings and the subsequent corrosion of steel.

THE FREQUENCY AND MECHANISM OF MICROBIAL DETERIORATION OF COATINGS

Failures of corrosion protective coatings are frequently experienced and also described in the literature. Where such failures cannot be traced back to chemical or physical causes one often talks about 'unsolved problems' (Funke, 1979). Some publications deal with coating deterioration and micro-organisms (Harris, 1966; Hedrick and Gilmartin, 1964; Becker and Gross, 1971): Micro-organisms have been found underneath blisters and debonded coating films on oil and gas pipelines, buried in the ground and in storage tanks for fuel oil. The authors have then mainly concentrated on the question of how the micro-organisms may have penetrated through the coating film in order to generate the deterioration between the coating and the metal surface.

In the course of our work large numbers of micro-organisms are frequently found on deteriorated coatings of different types and from various locations. Blisters may be filled with liquid and contain micro-organisms, but this is not always the case. Very often the inner surface of the blisters is sterile.

In our investigations, typical micro-organisms found in blisters did not exert a deteriogenic activity on new coatings, while the micro-organisms found on the outer surface often did so. We therefore postulate that the micro-organisms attack the coatings from outside, changing their chemical and physical properties in such a way that blistering and debonding may occur. This form of attack is common also with other materials, for instance plastics. When liquid and micro-organisms are found underneath the coating, they have probably penetrated through cracks caused by previous microbial deterioration.

Biocide-containing coatings, which cannot be attacked by micro-organisms, are often found to perform well in microbiologically active environments. Such investigations have been performed with inner coatings of fuel storage containers (Boggs, 1969; Kemp et al., 1966).

TESTING OF THE DURABILITY OF COATINGS

Surprisingly, no attention is paid to microbiological effects in the common short-term tests for the performance of organic coatings on steel. Neither do there seem to be any directives established for the microbiological testing of corrosion protective coatings.

In a comprehensive evaluation of short-term tests for coatings on structural steel (US National Bureau of Standards, 1981) micro-organisms are not mentioned at all. It is, however, concluded, that the test results are poorly reproducible and do not meet the need for relevant service life data.

Short-term tests are commonly performed in test chambers under conditions of humidity, temperature and time which are very favourable for the development of microbial populations. Obviously, different populations may develop in different chambers and at different iterations, and give different results.

We investigated the microbial flora on coated steel plates after removal from a common condensation test chamber. Large amounts of bacteria and fungi were found on most of the plates. These micro-organisms had been present during the test and surely acted on the coatings. No viable micro-organisms were present on some of the plates, particularly when the coating had performed well. This was, e.g., the case with a coal tar coating, applied on a zinc-containing primer. The corresponding sample without zinc-primer was blistered and populated with micro-organisms. Zinc is known to have a biocidal effect and, as mentioned before, added biocides often prevent failures of the coatings.

These observations indicate strongly that micro-organisms play a considerable role in the performance of coatings in common short-term tests. When they are ignored and not controlled, the results will certainly differ from test to test and be poorly reproducible.

MICROBIOLOGICAL TESTS FOR COATINGS

Experience up to now indicates that certain micro-organisms cause the described deterioration of corrosion protective coatings. Strains which have been isolated from deteriorated coatings are often the most aggressive ones for new coatings.

To select proper test organisms for the coatings in question, a number of cultures are screened with a spot test. Drops of melted agar media, containing spores of pure strains, are placed on the surface of coated steel plates. The plates are placed in a presterilized humid chamber and incubated. When the micro-organisms grow, some cause blistering and deterioration of the coating film underneath and around the agar drop.

Active fungal strains are then used in mixed culture for testing the microbiological resistivity of corrosion protective coatings. Inoculation and incubation of the test panels is performed according to common methods, as, for instance, described by Bravery et al., 1983. A certain amount of nutrients is added to the mixed spore suspension which is then sprayed on to the test panels. The speed of the attack can be enhanced by increasing the amount of nutrients, which may be advantageous when different formulations are to be screened within a short time.

The panels are placed in a humidity cabinet and incubated at ambient temperature. Fungal growth and blistering are evaluated from time to time, visually and under a microscope.

Corrosion protective paints often contain corrosion inhibitors. Some of these inhibitors are reported to give very good stability to coatings in short-term tests, on natural weathering and in practical use. It is striking that the effective inhibitors often are known biocides: zinc chromate, zinc borate, sodium benzoate, sodium nitrite, etc (Kossmann, 1985). This again indicates that microbial effects play an essential role in the deterioration of coatings.

We compared paints containing zinc borate and zinc phosphate as corrosion inhibiting pigments by testing their resistivity to microbiological attack, as described above. Very little microbial growth or blistering were observed on paints containing zinc borate, while the zinc phosphate containing paints became severely blistered within eight weeks.

Kossmann, 1985, in efforts to replace zinc chromate, which previously was used as an excellent corrosion inhibitor, also found zinc borate to give the best results amongst several other less biocidal compounds in a salt spray test and on natural weathering. The antimicrobial effect of borates is generally known.

CONCLUSION

Both in research, in testing and in practical cases of the deterioration of corrosion protective coatings microbiological effects are still greatly ignored. Because of the considerable economic consequences of coating deterioration and subsequent corrosion a large amount of work is spent on developing resistant coatings and test methods which could unambiguously predict the performance of the coatings in practical use.

According to our findings, more attention should be paid to microbiological effects. Such knowledge might greatly contribute to clearing the 'unsolved problems' which are mentioned in the literature. Microbiological tests should be included in the evaluation procedures for coatings on steel and should be made a rule before the coatings are applied in practice.

No short-term tests aimed at investigating the coatings' microbiological resistivity in marine environments are generally known today. These would involve the use of typical marine test organisms and possibly anaerobic test conditions. As there is increasing interest for protecting platform structures, subsea equipment and pipelines with coatings, the need for developing relevant test methods is obvious.

REFERENCES

Becker, H. and Gross, H. (1971). 'Beschädigung von Tankinnenbeschichtungen durch Mikroorganismen', *Farbe und Lack*, **77**, 533–539.

Boggs, W. A. (1969). 'Changes in aircraft design, fabrication and finish techniques and maintenance requirement dictated by microbe infested jet fuel sources', NACE, 25th Annual Conf., Paper 49, 15 pp.

Bravery, A. F., Barry, S. and Worley, W. (1983). 'An alternative method for testing the mould resistance of paint films', *J. Oil Colour Chem. Assoc.*, **66**, 39–43.

Funke, W. (1979). 'Corrosion tests for organic coatings — A review of their usefulness and limitations', *J. Oil Colour Chem. Assoc.*, **62**, 63–67.

Harris, J. O. (1966). 'Bacterial growth, unbonding of protective coatings and cathodic protection failure', NACE *Proceedings of the 2nd Int. Congress on Metallic Corrosion*, 358–363.

Hedrick, H. G. and Gilmartin, J. N. (1964). 'A detection study of microbiological penetration of aircraft fuel tank coatings', *Developments Ind. Microbiol.*, **6**, 124–132.

Kemp, H. T., Cooper, C. W. and Kell, R. M. (1966). 'Determining effectiveness of biocidal additives in coatings', *J. Paint Technol.* **38**, 363–367.

Kossmann, H. (1985). 'Wässrige lufttrocknende Korrosionsschutzanstriche', *Farbe und Lack*, **91**, 588–594.

Montle, J. F. (1971). 'Unexplained coatings phenomena with some partially baked answers', *Materials Protection and Performance*, **10**, 18–20.

Stranger-Johannessen, M. (1980). 'Aerobe sporenbildende Stäbchenbakterien als Deteriogene für kunststoffbasierte Korrosionsschutzanstriche und Fugenmassen', in *Biodeterioration* (Eds. T. A. Oxley, G. Becker, and D. Allsopp), pp. 143–147, Pitman, London.

Stranger-Johannessen, M. (1984). 'Fungal corrosion of the steel interior of a ship's holds', *Proceedings 6th Int. Biodetn. Symp.*, Washington, (in press).

Stranger-Johannessen, M. (1985). 'Mikrobiologiske skader på rustbeskyttende malinger og mikrobiologisk korrosjon', *Färg och Lack Scand.*, **31**, 125–129.

U.S. National Bureau of Standards. (1981). 'Short term evaluation procedures for coatings on structural steel', Technical note 1148.

RESERVOIR SOURING

B. N. Herbert

Shell Research Limited, Sittingbourne Research Centre, Sittingbourne, Kent ME9 8AG

SUMMARY

The generation of hydrogen sulphide in oil-bearing reservoirs during water injection can cause a number of problems such as the plugging of injection wells, souring of crude oil and gas production, corrosion of production facilities and toxicity to man. This hydrogen sulphide could be produced by both biological and geochemical processes. The biological mechanisms as they occur in reservoirs, mediated by sulphate-reducing bacteria, are beginning to be more clearly elucidated. Various geochemical processes are now being identified but are rather complex. However, their potential role in generating hydrogen sulphide in reservoirs is becoming better recognised.

This paper outlines how these processes could interrelate to produce hydrogen sulphide in reservoirs. A better understanding of these, together with the changes that occur in the reservoir environment during water flooding, should enable more successful and acceptable control strategies to be developed.

INTRODUCTION

The generation of hydrogen sulphide (souring) in oilfield reservoirs during water injection to maintain reservoir energies is usually thought to be due to the activities of sulphate-reducing bacteria (SRB). In offshore operations the injected water is usually seawater. Our understanding of the nutritional requirements of SRB (and how these can be satisfied) and the physico/chemical parameters that restrict their activities have been described (Herbert *et al.*, 1985). There are numerous obstacles to the growth of SRB in many reservoirs that are too hot and at too low a pH. In many instances the unflooded reservoir is initially an environment inimical to microbial activities except for those of thermophilic bacteria. There are, however, opportunities for more moderate environments developing as a result of water injection (e.g. cooling, pH rise). These changes may be difficult to quantify with any confidence because of the non-homogeneity of many reservoirs combined with insufficient detailed information of their total geology. The presence of variable permeabilities (e.g. thief zones), fracture planes and inclusion of different rock types (e.g. shale layers) can all affect water movements.

The discovery that substantial levels of short-chain fatty acids and ammonia can be present in many formation waters has identified a readily-available supply of carbon and nitrogen in forms that can be used directly by these strictly anaerobic bacteria as biosynthetic building blocks and as a source of energy (Herbert et al., 1985). Other organic compounds could also stimulate SRB indirectly if they are first utilized by fermentative bacteria and then pass on reducing power to the SRB. If oxygen enters the reservoir a similar process can occur but in this case hydrocarbons can become the primary carbon/energy supply.

Aerobic hydrocarbon-utilizing bacteria can pass on reducing power to SRB (in anaerobic regions) either directly or mediated by fermentative bacteria. Hydrogen is central to these processes and it is conceivable that there can be sufficient present in some reservoirs (Angino et al., 1984) to stimulate autotrophic sulphate reduction. Stetter and Gaag (1983) have shown that methanogenic bacteria can reduce sulphur to sulphide in the presence of hydrogen.

The idea that SRB are stimulated when injection and formation waters meet presumes that elements from both are required to produce the environment that supports good growth. For example, the formation water provides the organic nutrients; the injection water supplies the sulphate. There should be good mixing of the two waters if this effect is to be maximized. The injected water is usually expected to sweep through the reservoir as a front. Envisaged in this way, the only opportunities for mixing are limited to the interface region between two water types. Although this could involve a huge area it is still very small in relation to that of the overall reservoir. This region can also be expected to be too hot to support the activities of mesophilic SRB introduced via the injected water. In practice, however, pockets of formation water could be bypassed and thief zones will produce a 'tube' effect, so that opportunities for mixing in the cooled zones of the reservoir can exist.

PREDICTION OF RESERVOIR SOURING

Attempts to describe reservoir souring by a mathematical model are scarce. Ametov and Entov (1984) assumed that SRB activity was restricted to the the region of the injected well. They concluded that there was good agreement between experimental and theoretical data on the distribution of SRB and sulphates in relation to sulphide generation. This model ignores all the other factors that can play a part as discussed earlier, in particular, nutrient supply. One of the main obstacles to the prediction of the rate and extent of reservoir souring by SRB is the description of how and to what extent the physical and chemical environment develops as water injection proceeds.

The relationship between measured levels of available nutrients and the hydrogen sulphide observed in producing wells may not correlate with theoretical yields. The hydrogen sulphide in the gas phase may be at a concentration greatly in excess of maximum possible values obtainable from SRB activity even taking into account partition effects (see later for non-biogenic sulphide production). In other cases the sulphide level may be much less than expected. This could be due to nutrient imbalances but could also be as a result of hydrogen sulphide losses by migration to non-producing areas of the reservoir, by partition into the crude oil, by dilution, or by scavenging processes (e.g. by siderite). It may, therefore, be more realistic to concentrate on a relatively few important, easily determinable, parameters (Herbert *et al.*, 1985) in order to develop a model that predicts the worst case situation upon which engineering decisions can be taken. Using such an approach water injection systems that have an extremely low potential for microbial sulphide generation have been identified.

It is probable that two risk situations can be described. The first is a risk of high levels of hydrogen sulphide production (> 100 mg l^{-1} in the water phase) and assumes that the formation water supplies the nutrients and that mixing occurs. The second is a risk of chronic low levels of hydrogen sulphide production ($1-5$ mg l^{-1} in the water phase) in the region of the injection well. No mixing is required and the injection water supplies all the SRB requirements including nutrients at low levels.

Whatever happens (and it is possible that both processes occur) the hydrogen sulphide has to migrate to the producing wells before its level is measured — usually in the gas phase from a test separator. Hydrogen sulphide usually reaches the producing wells at breakthrough of injection water. There is, therefore, not only opportunity for the various concentration and dilution effects described above to take place but also a considerable and variable lag between the time the hydrogen sulphide is produced and the time it emerges. If reservoir characteristics are favourable, injection water breakthrough does not occur and the hydrogen sulphide is not seen.

SIGNIFICANCE OF HYDROGEN SULPHIDE LEVELS

The reasons for the undesirability of hydrogen sulphide in oilfield operations can vary. Apart from its toxicity to man (the threshold limit value is 10 ppm), it can cause serious corrosion to equipment by various mechanisms and when present in gas, above levels specified by customers, requires removal prior to distribution. Most of these problems can be overcome by suitable engineering and operational procedures. Some gas fields with 80% hydrogen sulphide content are in successful production. In fact the sulphur that results from the treatment processes is a commercial product. It is the uncertainty of whether or not a sweet reservoir will turn sour that can present difficulties in the selection of suitable sulphide-resistant materials. Various grades of steel have different resistances to the effects of hydrogen sulphide. Those most resistant are on the whole most expensive and hence considerable cost savings can be made if there is certainty that the reservoir will remain sweet. If a non-hydrogen sulphide-resistant steel is chosen then even low levels of hydrogen sulphide may produce problems such as hydrogen-induced cracking.

SULPHIDE GENERATION BY THERMOPHILIC SRB

It was believed until relatively recently that sulphate reduction could not occur chemically at temperatures below 250°C (see later) and hence any sulphide generation, even in hot reservoirs, must be biogenic. This view prevailed despite the lack of convincing evidence that there were SRB capable of activity above 75°C. Claims that SRB isolated from oilfield reservoirs are capable of activity at temperatures up to 105°C, though not widespread, are becoming more compelling even although the SRB have not been shown to be active in the reservoir. Over the past few years archaebacteria have been isolated from hot sulphur springs and volcanic areas that are involved in various aspects of the sulphur cycle. Of particular interest are those that can reduce molecular sulphur to sulphide under thermophilic conditions (Stetter et al., 1983).

It is possible that there are yet to be discovered thermophilic archaebacteria that can reduce sulphate and cannot be isolated by the conventional techniques used to isolate SRB from oilfields (Herbert and Gilbert, 1984). This is an exciting area of research which could reveal new unsuspected sulphide-generating bacteria that could revolutionize our concept of reservoir souring.

OTHER MECHANISMS OF SULPHIDE GENERATION

The appearance of hydrogen sulphide in produced fluids may be as a result of mechanisms other than SRB activity stimulated by water injection. Firstly, the sulphide may have been already present in pockets undetected during drilling. The number of wells drilled can only give a partial mosaic from which the overall geology is deduced from seismological data. Hydrogen sulphide has been detected (in some cases at concentrations in excess of 500 ppm in the gas) in appraisal wells drilled in many concession blocks throughout the North Sea. This sulphide is probably ancient and was probably produced by SRB during diagenesis. It is also possible that the sulphide was produced by a non-biological mechanism.

Although it was thought that non-biological sulphate reduction could not occur at temperatures below 250°C, a number of mechanisms have now been identified whereby hydrogen sulphide may be released or generated by a series of complex geochemical reactions in reservoirs at temperatures as low as 100°C or less.

Examples are:

a) Release from pyrite: for example during acidification procedures to improve injectivity (Bourgeois et al., 1979).
b) Thermochemical oxidation of hydrocarbons by anhydrite (Eliuk, 1984). Gypsum could also be a significant source.
c) Sulphur reduction by disproportionation (Belkin et al., 1985).
d) Thermochemical sulphate reduction in the presence of organic matter (Orr, 1977).
e) Reaction of pre-existing elemental sulphur with organic matter (Orr, 1977).
f) Thermal decomposition of organic sulphur compounds. This occurs during maturation (Orr, 1977).
g) Reaction of oxygen scavengers with steel (Whittingham and Hardy, 1985). This is limited to tubing, has been observed when ammonium bisulphite is used in excess and will probably not have a major role in reservoir souring. Ammonium bisulphite could also stimulate SRB activity by providing additional nitrogen and electron acceptor.

The reviews by Orr (1977) and Grinenko and Ivanov (1983) should be consulted for details of non-biological mechanisms of sulphide generation.

Elemental sulphur is probably the most stable form of sulphur and the more reduced or oxidized forms have varying degrees of increased instability.

It is conceivable that a number of geochemical processes are occurring in a reservoir concurrently with biological sulphate reduction. Biologically generated sulphide can be distinguished from geochemical sulphide by its sulphur isotope ratio. However, if both processes are taking place, or if one of the mechanisms is totally swamped, differences could be blurred to such a degree that sensible interpretation is confounded. Nevertheless, it is strongly recommended that sulphur isotope ratio data on the sulphide produced are obtained as such information could strongly influence any control strategies.

CONTROL OF SRB ACTIVITIES IN RESERVOIRS

The information gained from an initial study of the likely souring mechanisms in a particular reservoir can be of great assistance firstly in deciding whether bacterial sulphate reduction is likely to be of significance and secondly in devising a control strategy. Even if it is decided not to use biocide treatment at the outset of water injection it would be prudent to have sufficient data available on acceptable biocides (in terms of compatibility with other production chemicals, environmental aspects, toxicity, etc.) so that one can proceed with a treatment when the time is judged right. There is considerable variation between oil companies as to whether they elect to treat with biocides at the initiation of water injection or await developments. In some cases the injection water is only treated with biocide when the injection wells are shut in as it is considered that this is the greatest period of risk of significant SRB activity in this region.

It was outlined earlier that the prediction of the maximum level of hydrogen sulphide achievable by SRB activity in a particular oilfield can lead to proper material selection which could obviate the need for biocide treatment. If plugging of the reservoir occurs this can be remedied by various treatments including acidification. This acceptance of hydrogen sulphide carries the penalty of increased requirement for protection of staff and treatment processes. However, there are various known technologies for removal of hydrogen sulphide from gas (e.g. Sulferox).

The knowledge that the sulphide has not been produced by current SRB activity (sulphur isotope analysis) would prevent the unnecessary use of biocides. Detailed information and understanding of how the changing reservoir environment during water flooding relates to our knowledge of SRB physiology could point to treatment procedures that avoid the use of biocides. These include, for example, prevention of reservoir cooling by heating injection water, pH control and redox control at values which prevent SRB activity but do not increase the risk of tubing corrosion. The use of cathodic protection (-900 mV) could be beneficial in preventing SRB activity on the surface of injection well tubing.

The quality of the injection water is usually defined in terms of total dissolved solids and filterability (Mitchell, 1981). No account is taken of dissolved organic matter content which, as stated earlier, can result in chronic but significant sulphide generation in the region of the injection well. In some cases the production chemicals themselves could stimulate SRB activity (e.g. ammonium bisulphite) and where possible effective alternatives should be used. Various biodegradeable organic molecules (e.g. biopolymers) are being considered for enhanced oil recovery and these could, without adequate microbial control, stimulate SRB activity. In addition, the presence of hydrogen sulphide can result in degradation of polyacrylamides.

Whilst all means to control SRB without the use of biocides should be considered and investigated, there is no doubt that in many instances there will be no choice but to use biocides. They have been used successfully for the control of SRB in surface facilities. However, their efficacy within reservoirs is less clear. The use of squeeze treatments has been claimed to be successful although often transitory. In such cases it can be difficult to distinguish between inhibition of SRB activity and scavenging of hydrogen sulphide as would occur with aldehydes. Data on the stability and mobility of biocides in reservoir environments are scanty. Chemical suppliers can often supply data on biocide stability with respect to temperature and pH and these may be used to make a judgement on likely stability. The mobility of a biocide through a reservoir will be a function of stability and losses through adsorption processes. Chromatographic effects in reservoirs could separate components of biocide formulations resulting in precipitation of the active ingredient(s). Loss of biocide activity due to instability may be compensated for, in part, by increased activity as temperature and pressure increase (Stott and Herbert, 1986).

Results from some oil operations suggest that some biocides can migrate through reservoirs. Various complex molecules can migrate (as indeed oil does) through the reservoir and the successful use of enhanced oil recovery chemicals depends upon this effect. Any biocides that are used as treatments in these cases will be expected to co-migrate through the reservoir with the chemical. There is scope for much more research in this area, not only on available biocides but also in the search for new tailor-made molecules. Whilst there are many hundreds of biocides currently on offer there are only a few (possibly only two or three) that can be considered for use to control SRB activity in a particular water injection programme.

In a situation in which the only SRB activity occurs in the cooler part of a reservoir, an ideal biocide would be one that is stable at the temperatures favouring SRB activity but breaks down as the temperature rises and takes over as the controlling factor. This could prevent any subsequent environmental damage at discharge. The possibility that the successful use of biocide treatment in a reservoir can cause subsequent problems cannot be excluded and needs to be carefully considered before a treatment programme is initiated. Apart from environmental problems of discharges into the open sea or rivers, other processes such as biotreaters or sludge treatments (in soil) could be adversely affected. One way to overcome such environmental problems could be, where practicable, to reinject produced water.

Finally, there has been interest recently in the use of ionizing radiations to control SRB in injection facilities. The use of ultraviolet radiation has been suggested to kill SRB in injection waters (Ege et al., 1985) whilst gamma radiation has been put forward as a means of killing SRB at the bottom of the well bores as the water enters the reservoir (Agaev et al., 1985). The former claim seems to have most merit on technical grounds but could only be used to reduce a bacterial challenge to a system as there is no residual activity and 100% kill cannot be guaranteed. In addition there may still be a need for a chemical biocide treatment downstream of the ultraviolet unit. When 'clean' seawater from temperate waters is used, having an SRB population of less than 1 cell l^{-1}, any further reduction may not be justified. However, waters containing greater SRB numbers (e.g. tropical lake waters, some aquifers, reinjected produced waters) may benefit from such a treatment by reducing the bacterial challenge and so making chemical biocide treatment more effective.

REFERENCES

Agaev, N. M., Smorodin, A. E. and Guseinov, M. M. (1985). 'Influence of gamma radiation on the metabolic activity of sulphate-reducing bacteria', *Prot. Met.*, **21** (1), 111–113.

Ametov, A. M. and Entov, B. M. (1984). 'Determination of parameters of models of biogenic sulphate reduction in oil beds', *Izv. Vyssh, Uchebn. Zaved., Neft. Gaz.*, **27** (7), 31–35.

Angino, E. A., Coveney, R. M., Geobel, E. D., Zeller, E. J. and Dreschhoff, G. (1984). 'Hydrogen and nitrogen — origin, distribution, and abundance, a follow up', *Oil and Gas J.*, Dec. 3, 142–146.

Belkin, S., Wirsen, C. O. and Jannasch, H. W. (1985). 'Biological and abiological sulphur reduction at high temperatures', *Appl. Environ. Microbiol.*, **49** (5), 1057–1061.

Bourgeois, J. P., Aupaix, N., Bloise, R. and Millet, J. L. (1979). 'Proposition d'explication de la formation d'hydrogene sulfure dans les stockages souterrains de gaz natural par Réduction des sulfures mineraux de la rocke magasin', *Revue de l'Institut Français du Petrole*, **34** (3), 371–386.

Ege, S. L., Houghton, C. J. and Tucker, B. M. (1985). 'UV — Control of SRB in injection water', *Conference Proceedings of U.K. Corrosion '85* Harrogate, November 4–6, 1985, 243–258.

Eliuk , S. E. (1984). 'A hypothesis for the origin of hydrogen sulphide in Devonian Crossfield Member Dolomite, Wabamun Formation, Alberta, Canada', *Conference Proceedings. Canadian Soc. Petrol. Geol. Carbonate in Sub Surface and Outcrop Core*. October 18–19, 1984, 245–289.

Grinenko, V. A., and Ivanov, M. V. (1983). 'Principal reactions of the global biogeochemical cycle of sulphur', in *The Global Biogeochemical Sulphur Cycle* (Eds. Ivanov, M. V., and Freney, J. R.) Wiley, Chichester, 1–23.

Herbert, B. N. and Gilbert, P. D. (1984). 'Isolation and growth of sulphate-reducing bacteria', in *Microbiological Methods for Environmental Microbiology*, Society for Applied Bacteriology Technical Series 19, Academic Press, London, 235–258.

Herbert, B. N., Gilbert, P. D., Stockdale, H. and Watkinson, R. J. (1985) 'Factors controlling the activity of sulphate-reducing bacteria in reservoirs during water injection', *Conference Proceedings. Offshore Europe '85, Aberdeen*. September 10–13, 1985. Society for Petroleum Engineers SPE13978/10.

Mitchell, R. W. and Finch, E. N. (1981). 'Water quality aspects of North Sea injection water', *J. Petr. Technol.*, **33** (6), 1141–1152.

Orr, W. L. (1977) 'Geologic and chemical controls on the distribution of hydrogen sulphide in natural gas', *Adv. Org. Geochemistry*. (Eds. Campos, R. and Goni, J.) 571–597. Conference Proceedings. Madrid 1975. Empresa Nac. Adaro. Invest. Minera, SA. Madrid.

Stetter, K. O., and Gaag, G. (1983). 'Reduction of molecular sulphur by methanogenic bacteria', *Nature, London*. **305**, 309–311.

Stetter, K. O., Konig, H. and Stackebrandt, E. (1983). *'Pyridictium* gen. nov., a new genus of submarine disc-shaped sulphur-reducing Archaebacteria growing optimally at 105°C', *System. Appl. Microbiol.*, **4**, 535–551.

Stott, J. and Herbert, B. N. (1986). 'The effect of pressure and temperature on sulphate-reducing bacteria and the action of biocides in oilfield water injection systems', *J. Appl. Bact.* **61**, 57–66.

Whittingham, K. P. and Hardy, J. A. (1985). 'Microbial corrosion control in water injection systems', *Conference Proceedings. U.K. Corrosion* '85, Harrogate, November 4–6, 1985, 271–280.

MICROBIAL ENHANCED OIL RECOVERY TECHNIQUES AND OFFSHORE OIL PRODUCTION

Jean L. Shennan and I. Vance

BP Research Centre, Sunbury on Thames, U.K.

INTRODUCTION

Microbial systems for enhanced oil recovery have been used for many years, mainly in on-shore fields in the United States and in Eastern Europe, and successful enhancement of oil production has been claimed. Accordingly, microbial enhanced oil recovery (MEOR) may, at first glance, appear to be an option to be considered along with other EOR methods being assessed for off-shore operations in an attempt to win additional oil. This paper reviews MEOR as currently practised in shallow, stripper wells and raises the practical, technical and economic questions which appear to preclude its use, at our present state of knowledge, as an EOR mechanism for use in UK continental shelf (UKCS) fields.

Microbial EOR is defined as the use of micro-organisms to good effect within the reservoir, involving injection of selected micro-organisms and the stimulation of growth *in situ* so that the presence and activity of cells leads to the increased production of crude oil.

The production *ex situ* of a particular class of microbial metabolites, biopolymers, under factory conditions followed by addition to the reservoir in the conventional way with the water flood is discussed by Sutherland (1986).

What is Enhanced Oil Recovery?

Conventional primary and secondary recovery techniques for oilfield exploitation, if used efficiently, extract, on average, 35–45% of the original oil-in-place. The proportion recovered varies between fields and between reservoirs, depending on the viscosity of the crude, the characteristics of the reservoir rock and the technology and economics of production.

Primary oil recovery relies on the energy stored within the reservoir to force crude oil through the porous rock formation to the production well.

73

Mechanisms include simple expansion of oil as reservoir pressure drops, expansion of the gas cap above the oil-bearing region, expansion of gas escaping from solution in the oil, expansion of water held under pressure in aquifers associated with the oil zone and gravity drainage of oil in steeply dipping reservoirs. Secondary recovery methods are based on direct displacement of crude oil by water pumped into the reservoir or gas injected into the gas cap.

During secondary recovery, significant quantities of oil are left behind in areas of the reservoir which are not swept by the water or gas floods or remain trapped in capillary pores of the formation rock. Enhanced (or tertiary) recovery methods are intended to extract more of this oil in an economic timescale. The techniques can be divided into four broad groups: chemical, miscible gas, thermal and 'exotic', in which last category lies *in situ* microbial enhanced oil recovery.

Although all manner of oil fields exist in many combinations of size, depth, reservoir geology and crude oil type, in the context of this review two main categories are considered: large, deep, hot, often off-shore fields and small, shallow, cool 'stripper' wells on-shore. Deep, hot fields in the UKCS have been on secondary production for less than 15 years and tertiary oil recovery techniques have not yet been implemented in these regions.

Stripper wells (producing less than 10 bbl oil/day), with their low production costs, can, under appropriate tax regimes, continue producing economically only a few barrels of oil and a great deal of water each day for many years. Although actual figures are hard to come by, it may be expected that large, deep fields, often producing from off-shore platforms, will become uneconomic to operate once significant levels of unavoidable water breakthrough occur. This reduces the rate at which crude oil is produced and, hence, the financial return against the high running costs.

In North Sea fields, with a generally higher than average recovery by conventional production, the EOR target may be smaller than in some other oilfields around the world. However, if an additional 5% of the original oil could perhaps be obtained by EOR techniques, this would represent some 180 million barrels of oil in the Forties field alone. This is equivalent to the reserves of an additional medium-sized North Sea reservoir.

Although the value of the residual oil may be high, there are considerable technical difficulties involved with its recovery from offshore oil operations. Surfactant EOR, for instance, is likely, for a number of reasons, to be conducted by large scale injection of a low concentration (e.g. 1.5% v/v) of surfactant in seawater. The low concentration is dictated by logistics of chemical carriage and storage on platforms and the possibility that small volume slug additions could become dispersed in a field with wide well spacings.

While the principal limitations to the widespread adoption of EOR techniques are technical ones, it has to be stressed that functional success is only part of the EOR goal. To the field operator, heavy capital investment and long payout times are unwelcome late in the life of a dwindling asset and so economic feasibility is crucial (Grist, 1983). This may be affected by changes in the price of crude oil and distorted by national taxation policies which are, again, dependent upon the planned timescale for EOR applications. In addition to technical and economic constraints, the potential for most EOR processes is also affected by logistics, especially for fields off-shore or in otherwise hostile environments.

"Conventional" EOR Techniques

Microbial EOR has to bear comparison with other, more thoroughly researched methods of enhanced oil recovery (Grist, 1983; Hann & McGillivray, 1985). Thermal methods are well established and have been shown to be economically viable in shallow reservoirs containing heavy viscous oil which would otherwise be unrecoverable. Gas injection, to generate a light solvent in the reservoir, uses associated petroleum gases (in areas where there is no market for this product). Where a cheap source of CO_2 gas is available, CO_2 flooding is now regarded as an established technique and is used on a large scale in the USA. Chemical flooding for EOR, i.e. the addition of surfactants, polymers or alkalis to the injection water, has also been actively researched by most of the large oil companies and used, to some extent, in on-shore fields. Economic restraints here require that the chemicals be low cost or highly efficient at oil displacement.

These EOR methods are all fairly well understood through laboratory and field testing and their use in the field can be modelled in computer simulations when EOR schemes are proposed for specific reservoirs.

The Case for *In Situ* MEOR

Chemical EOR modifies the water flood from the point of injection and so will exert its effect wherever the water flood sweeps. However, loss of chemical adsorbed to rock surfaces, or other losses due to the physical and chemical conditions of the reservoir will occur throughout the formation, reducing the amount of active agent available for the intended EOR mechanism. It has been argued (Moses & Springham, 1982) that some EOR benefit could be gained if oil mobilizing agents were continuously produced at the water/oil interface by microbial action. A continual loss of material by adsorption would be acceptable if replacement quantities were being produced fast enough by micro-organisms. The amount of metabolite needed is claimed to be less than in a chemical EOR flood as the active agent would be produced at the site of maximum effect. Under optimal conditions (which require definition), metabolites would also be produced over extended periods of time in the reservoir and at locations distant from the injection well.

IN SITU MICROBIAL ENHANCED OIL RECOVERY

General Process Concepts for *In Situ* MEOR

Microbial EOR has been the topic of a number of International Symposia and review articles (Moses & Springham, 1982; Donaldson & Bennett-Clark, 1983; Finnerty & Singer, 1983; Zajic *et al.*, 1983; Springham, 1984; Zajic & Donaldson, 1985; Shennan & Levi, 1986). The general principles are similar for most schemes proposed or tested.

Micro-organisms, selected to be active in the anaerobic environmental conditions of the reservoir, are introduced into the formation with the water flood. The cells may be added either as an active culture or in a dormant state. Provided that rock permeabilities are suitable and the cells do not block the formation near the well bore and hence impair injectivity (Bubela & McKay, 1985), the inoculum will be swept into the formation with the water flood until the cells come to rest in narrow pore spaces, attach to rock surfaces of pore throats or settle at oil/water interfaces as the force of the injection water flow is dissipated at some distance from the injection well.

A nutrient supply, usually molasses, is injected with or after the cells to support growth of these anaerobic bacteria and carbohydrate ferment-ative metabolic pathways relied upon to produce active metabolite(s). Adequate levels of mineral nutrients needed for growth are likely to be found in the injection water, in the supplied carbohydrate substrate or are already present in the formation.

Most MEOR trials on-shore have then included a period of well 'shut-in' with cessation of the water flood for periods of weeks to months to allow for development of the culture underground. On resumption of the water flood, oil output is assessed for enhancement.

In Situ MEOR Mechanisms

Microbial mechanisms of oil release, as assessed in the laboratory, are usually ill-defined and difficult to examine in detail. Several mechanisms which could lead to crude oil release from the formation include: generation of CO_2 (increasing oil volume by solution and reducing viscosity); synthesis of organic acid metabolites which dissolve carbonate rocks; production of biopolymers which increase the viscosity of the water flood and of surfactants which reduce oil-water interfacial tension and release capillary entrapped oil. More than one mechanism is likely to be operating in any given MEOR trial (which may be considered an advantage *versus* 'conventional' oil recovery from stripper wells).

Oil-water emulsification is an important property of micro-organisms capable of assimilating liquid hydrocarbons, as a prerequisite for the transfer of the immiscible substrate into the cell. Many of the usual lipid metabolites of cells are potentially surface active, but growth in the presence of hydrocarbons appears to stimulate both the production and release of extracellular surfactant chemicals (Kosaric *et al.*, 1984; Zajic & Seffens, 1984). This effect is also observed when the surfactant-producing organisms are growing on carbohydrates (Guerra-Santos *et al.*, 1984; Chakrabarty, 1985; Duvnjak & Kosaric, 1985).

Another proposed mechanism, completely different in concept and application, is the selective plugging of high permeability thief zones (water channels) in a reservoir by bacterial cells and/or their polymeric metabolites. This will be discussed in a later section.

Transport Problems of Micro-organisms in Porous Media

The introduction of viable cells into the formation for *in situ* MEOR introduces yet another component into the already complex multi-phase system (rock/water/oil) of the reservoir. While, undoubtedly, the crucial effects of *in situ* MEOR will take place within the micro-habitat of the cell/oil/water/solid complex of interfaces, the propagation of these effects throughout the reservoir will be influenced by the movement of oil-mobilizing extracellular bacterial products and the mobility of bacteria.

In dynamic systems, bacterial cells will approach rock surfaces in the fluid stream by laminar flow aided, perhaps, by their own motility. Taxis or trophism may even occur, attracting cells towards the hydrocarbon contact surface. The physico-chemical properties of the reservoir rock, including the number of available sites for retention on the rock surface, the interactive forces between cell and rock, and specific means of adhesion will all affect the rate and degree of attachment of bacterial cells onto solid surfaces.

The rate of bacterial penetration in porous rocks can be assumed to be affected by the porosity, permeability, mineral composition and wettability of the rock minerals. The hydrodynamics of the flood system will also influence cell transport through the reservoir rocks.

Laboratory studies have improved understanding of the behaviour of viable microbial cells suspended in a liquid injected into porous rock and in rock saturated with non-flowing liquid (Jang *et al.*, 1983; Bubela, 1984; Jang & Yen, 1985; Jenneman *et al.*, 1985; Torbati *et al.*, 1985; Dow & Quigley, 1986).

Even in static systems, bacterial penetration rates in the range of 0.01 to 0.11 metres per day have been observed in Berea sandstone with permeabilities of 70.7 to 182 mDarcies. Both motile and non-motile species of bacteria traversed sandstone cores, although the penetration rate of the non-motile species was up to 8 times slower (Jenneman *et al.*, 1985). In sandstone saturated with nutrient-containing brine, the rate of penetration of a motile *Bacillus* sp. was found to decrease in permeabilities below 100 mDarcies but was independent of permeability above this value (Jenneman *et al.*, 1985).

In Situ MEOR Field Trials

In situ MEOR has so far only been reported in small stripper wells (less than 10 bbl/day oil production). As a consequence, the EOR information obtained relates to largely watered-out fields, not those in which a significant amount of oil is still being recovered, as might be the case should EOR be attempted in off-shore fields.

The reasons for confining trials of MEOR to stripper wells are obvious, and are related mainly to the small financial loss incurred should the trial be unsuccessful or the treatment in fact be detrimental to production. In other words, a modest real increase in oil production from an almost watered-out field will represent a high percentage increase in yield. Should treatment be unsuccessful, a decrease in production will not drastically decrease the productive life expectancy of such a field; it will have been near to closure anyway.

Small, shallow, on-shore fields are numerous in some regions of the world, for instance in the USA, USSR and Eastern Europe, and it is from these areas that field trial results have become available.

Bacterial treatment of stripper wells, mostly involving single well treatments with widely varying results, has been reviewed by several authors (Moses & Springham, 1982; Yarborough & Coty, 1983; Zhang & Qin, 1983; Springham, 1984; Janshekar, 1985). Hitzman (1983) compiled data for over 200 wells, typically producing 1–2 bbl/day of oil, treated by *in situ* MEOR techniques since 1954. Despite the mass of data tabulated in his review, only gross changes can be detected and specific effects due to different mechanisms cannot be discerned. Lazar (1983a) has reviewed MEOR practices carried out in Romania since 1971. Analysis of produced waters for numbers and types of bacteria indicated that inoculated bacteria did spread through the formation and reappeared in the produced fluids but that contaminating bacteria in the formation were also stimulated into growth. Often, however, it was impossible to prove whether cessation and later resumption of water flooding of the well or the MEOR treatment had enhanced the oil yield.

Whether MEOR due to *in situ* metabolite production can be effective without stopping the water flood has not been assessed in the field.

The conflict between activity attributable to enhanced treatment methods and that resulting from mere perturbation of the water flood regime, has also been reported for MEOR in a 16-well field test carried out in Last Chance, Colorado (Parkinson, 1985). The microbial treatment was preceded by the injection of surfactant (to prevent plugging) which further confused the outcome of the experiment.

Fields tests have been carried out in four states of the USA by a team from Hardin-Simmons University (Petzet & Williams, 1986). Since 1982 they have injected 80 wells and obtained increase in pressure through microbial production of CO_2 in 64 of these wells. More than 40 of the wells injected showed some increase in production, although this took from 2 weeks to several months to appear, lasted for a few weeks to more than 9 months and usually yielded only a few barrels per day over the original level.

Other trials are reported (Petzet & Williams, 1986) to be under way or planned in the USA.

Available analyses from on-shore MEOR field trials do not provide the quality and quantity of data needed to assess the MEOR mechanisms at work. Assessments can only be made by extrapolating from known facts on extracellular production of microbial metabolites *in vitro* and from the ecological and metabolic behaviour of micro-organisms under simulated reservoir conditions.

COULD *IN SITU* MEOR WORK IN UKCS RESERVOIRS?

Evidence that *in situ* MEOR appears to work in shallow reservoirs (although by uncertain mechanisms) has encouraged at least an evaluation of this EOR process using micro-organisms in comparison with conventional EOR systems which may be considered for use in off-shore oil fields.

Table 1   Characteristics of some North Sea reservoirs

Oilfield	Temperature (°C)	Initial Pressure (bar)	Salinity (%TDS)
Piper	80	240	7.5
Brent	93	410	1.0
Forties	96	224	10.2
Thistle	102	410	1.3
Ninian	102	450	2.0
Magnus	115	475	1.1

The major constraints upon microbial metabolism for off-shore MEOR applications have been assumed to be the initially high temperatures of these deep reservoirs (Table 1). The injection of sea water for secondary oil recovery will lead to significant cooling of the reservoir in the immediate vicinity of injection wells. The injected water front, however, will be at a considerable distance from the well bottom and is unlikely to be significantly cooler than the formation through which it is passing. In designing a hypothetical *in situ* MEOR process, therefore, it has to be assumed that, in order to be effective in mobilizing the target residual crude oil, the bulk of which is at some distance from the injector well, micro-organisms will have to be active throughout the prevailing temperature profile of the reservoir. The quantitative distribution of fermentation end-products resulting from microbial metabolism of added substrate is also expected to be sensitive to changing temperatures across the flooded reservoir (see below).

A search for suitable MEOR cultures, able to grow and produce effective metabolites in the predicted reservoir environment is, therefore, required before any technical assessment can be made of *in situ* MEOR for off-shore use. Operational constraints will be discussed later.

THE SELECTION OF ORGANISMS FOR *IN SITU* MEOR

Choice of Organisms

The constraints on the selection of suitable strains for *in situ* MEOR are several and are largely imposed by the prevailing conditions in reservoirs. Some factors are common to all oil-fields, e.g. absence of free oxygen, while other factors, such as temperature, will vary between fields and, in some cases, will be at or beyond the limits for known microbial activity.

Downhole conditions are anoxic, as operators go to great lengths to deoxygenate flood water (to less than 50 ppb) to guard against corrosion of well casings and downhole equipment. Consequently, crude oil fractions cannot be exploited because a growth substrate as the first step of the hydrocarbon breakdown pathway is oxidative, involving the enzymatic insertion of molecular oxygen into the hydrocarbon molecule (Gibson, 1984).

While some degree of biological conversion of crude oil appears to be possible in the reservoir over a geological timescale, as described by Connan (1986), mechanisms of anaerobic attack on hydrocarbons at rates feasible for EOR must involve an inorganic oxidant. Authentic reports in the literature of anaerobic utilization of hydrocarbons are based on nitrate- or sulphate-driven respiration (e.g. Springham, 1982; Britton, 1984; Sleat & Robson, 1984; Young, 1984; Jack et al., 1985). However, nitrates would not be introduced downhole for reasons of corrosion prevention. Similarly, much effort is devoted in oilfield practice to preventing the production of biogenic sulphides as a result of dissimilatory sulphate reduction by bacteria. The serious production difficulties caused by sulphides have been discussed by Herbert (1986).

MEOR organisms must, therefore, be selected from obligate or facultative anaerobic species and be able to utilize a readily available and inexpensive carbohydrate substrate. However, the thermal degradation of carbohydrate substrates experienced in laboratory experiments with thermophilic bacteria (Wiegel et al., 1979) may restrict their use in deep hot reservoirs.

Significant concentrations of short chain fatty acid anions, notably acetate and propionate, have been observed in formation waters from oil and gas fields in California and Texas (Carothers & Kharaka, 1978). These could, in theory, act as carbon and energy sources for a modest amount of anaerobic metabolic activity.

The selected MEOR organism (or population of organisms) must produce from the non-hydrocarbon substrate an agent (or mixture of agents) which will mobilize crude oil in the porous rock matrix of the reservoir. These active agents will be end-products of fermentative metabolism, such as CO_2, organic acids or solvents, or extracellular polymeric products — surfactants and biopolymers.

Organisms for in situ MEOR in Deep, Hot Reservoirs

Organisms suitable for in situ MEOR in high temperature, off-shore reservoirs able to tolerate the constraints outlined above are not immediately identifiable. Therefore, before any assessment can be made for the potential for in situ MEOR or analysis of the mechanisms that might be effective, a search has to be made for organisms capable of growth and oil mobilization, without plugging, under the extreme temperature environments of these fields.

In recent years, the rapid expansion of studies into microbial existence in extreme environments has led to the discovery of new strains of bacteria growing in temperature conditions previously thought unlikely to support life (Brock, 1985; Deming, 1986). As these organisms use as yet ill-understood strategies of survival, the hope that organisms suitable for use in extreme reservoir conditions might exist was rekindled for a brief period. This possibility was encouraged by claims made for the isolation from submarine hydrothermal vents of bacteria capable of growth up to 250°C at reservoir pressures (Baross & Deming, 1983), evidence for which is now thought to be artefactual (Trent et al., 1984; White, 1984). However, extremely thermophilic anaerobes, growing at 80°C to 110°C have been

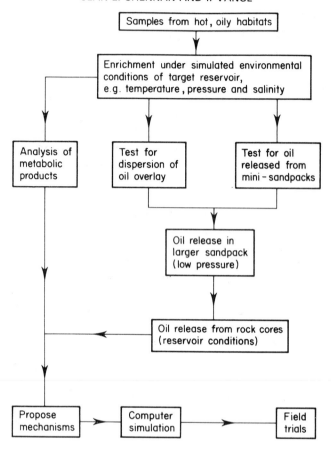

FIG 1   Strategy for screening bacterial cultures with potential for MEOR.

isolated and grown in the laboratory. These organisms, all archaebacteria, fall into several nutritional categories, living autotrophically on sulphur, hydrogen and carbon dioxide, by methanogenesis, heterotrophically on organic substrates by sulphur respiration or, in a few cases, by anaerobic fermentation (Stetter, 1984).

SCREENING STRATEGY FOR *IN SITU* MEOR CULTURES

Sources of MEOR Organisms

The most successful cultures used for *in situ* MEOR field trials in cool, stripper wells, appear to have been mixed populations usually obtained from organically rich wastes and from anaerobic oil-polluted environments (Lazar, 1983b; Bubela & McKay, 1985).

In the search for potential MEOR cultures for high temperature fields, hot 'oily' environments (like oil/water separators on North Sea production platforms) were screened for anaerobic, carbohydrate-fermenting thermophiles tolerant of pressure and moderate salinity. A generalized screening regime for MEOR organisms illustrating the experimental approach adopted in the authors' laboratory (modified from Levi et al., 1985) is shown in Figure 1.

Enrichment Under Simulated Field Conditions

The need to enrich samples and test cultures under conditions of high temperature and high pressure has required the development of appropriate test equipment. In the authors' laboratory anaerobic bacteria have been cultured in serum vials or glass syringes with stoppered nozzles within cylindrical pressure vessels filled with silicone fluid. Pressure was applied by a hand or air-driven pump and the pressure vessels were heated by immersion in an oil bath. Bacterial growth was observed at combinations of temperature and pressure as high as $75^\circ C$ at 400 bar.

More sophisticated apparatus can be constructed to allow subsampling for kinetic growth studies during high temperature/pressure incubations (Yayanos et al., 1984; Vance & Richmond, 1985).

Screening Cultures for Growth and Production of Effective Metabolites

In addition to the combined temperature/pressure apparatus described above, the effect of pressure alone on microbial physiology and morphology has been studied using a variety of specially devised pressure cells (Zobell & Oppenheimer, 1950; Marquis, 1976) or modified French pressure cells (Vance & Hunt, 1985).

Conventional techniques of microbial culture and chemical analysis can be used to assess growth and production of selected metabolites under conditions of reservoir temperature and salinity (Vance & Richmond, 1985).

Screening for Oil Release from Sandpacks

The screening of suitable cultures for oil mobilization by microbial mechanisms can be carried out in sandpack and reservoir core apparatus in a similar fashion to that used for screening EOR chemicals. In the authors' laboratory, a rapid screen for crude oil release (at atmospheric pressures and temperatures up to $70^\circ C$) was carried out using 'mini-sandpacks', made from standard laboratory glassware, with a means for collecting the produced oil (Levi et al., 1985). Thermophilic, anaerobic bacterial cultures were then evaluated for oil mobilization in pressurised sandpacks constructed from HPLC columns.

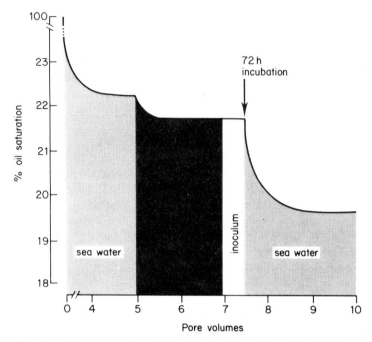

FIG 2   Oil release from a sandpack. Sandpack: dimensions 50 × 2.25 cm diameter, packed with Forties separator sand (50–150 um particle size range), porosity 40%, permeability 2 Darcies. Experiment: The sandpack was saturated with synthetic Forties produced water, then flooded with Forties crude oil and finally flooded with seawater until no further oil was released; a residual oil saturation $(SOR)_w$ of 22.1%. The sandpack was then flooded with 2 pore volumes of growth medium giving a small additional release of crude oil to $(SOR)_w = 21.7\%$. The bacterial inoculum (0.6 pore volumes) was washed into the sandpack (no further release of crude oil from the inoculation step) and the sandpack was incubated (60°C, 4 bar(ga)) for 72 hours. On resumption of the seawater flood a modest release of crude oil was obtained $(SOR)_w = 19.62\%$.

More elaborate sandpacks and, ultimately, devices using rock cores held at reservoir temperature and pressure, would be required for the complete laboratory evaluation of microbiological techniques for enhancing oil recovery.

Results of Screening Exercise

In the authors' laboratory, the strategy described produced several isolates capable of modest crude oil mobilization from laboratory sandpacks with some decrease observed in the 22% residual oil saturation (post water flood) (Figure 2). However, the upper temperature limit for effective cultures was found to be around 75°C. No growth or metabolite (ethanol or volatile fatty acids) production was detected at 80°C for these oil-mobilizing cultures.

Table 2   Effect of temperature upon glucose fermentation products in cultures of an anaerobic thermophilic bacterium B679 isolated from an oil separator on a North Sea oil production platform

Temp °C	Glucose* fermented mM	Fermentation End-products				Yield mole/mole glucose			Carbon Recovery %
		Ethanol mM	Acetic acid mM	Lactic acid mM	CO_2[†] mM	Ethanol	Acetic	Lactic	
70	29.9	21.1	2.7	40.0	23.8	0.7	0.09	1.3	107
75	93.5	162.0	3.2	23.3	165.2	1.7	0.03	0.25	101

*Initial concentration 94.4 mM.
[†]CO_2 estimated; 1 mole CO_2 per mole ethanol or acetic acid.

Table 3   Effect of pressure and temperature on growth and ethanol production in an enrichment culture of anaerobic thermophilic bacteria isolated from an oil separator on a North Sea oil production platform

Incubation conditions	Increase in absorbance at 600 nm	Increase in ethanol concentration g litre^{-1}
@ 70°C for 22 hours		
atmospheric	1.10	2.49
100 bar (ga)	0.27	0.81
200 bar (ga)	0.08	0.23
300 bar (ga)	0.04	0.13
400 bar (ga)	0.03	0.09
@ 75°C for 71 hours		
atmospheric	0.48	2.19
100 bar (ga)	0.67	3.30
200 bar (ga)	0.51	1.77
300 bar (ga)	0.03	0.16
400 bar (ga)	0.02	0.09
@ 80°C for 48 hours		
atmospheric	0.01	0
100 bar (ga)	0.01	0
200 bar (ga)	0.01	0
300 bar (ga)	0.01	0
400 bar (ga)	0.01	0

For some cultures, a test at atmospheric pressure showed that incubation temperature influenced the balance of fermentation products obtained from glucose (Table 2). An increase of 5 degrees Celsius (70°–75°C) altered the major product from lactic acid to ethanol, a swing which could have profound consequences on an MEOR process theoretically based on one or other of these metabolites.

Pressure effects on the growth of these isolates appeared to be less critical than temperature effects. Inhibition of growth was found above 200 bar (Table 3) although selected isolates grew (but with a longer lag phase and doubling time) up to 400 bar at 75°C. For *in situ* MEOR, pressure effects on cell morphology may be more serious. Strains which grow as short rods at atmospheric pressure may form long filaments when grown at elevated pressures (Zobell & Oppenheimer, 1950; Zobell & Cobet, 1962).

It is apparent, therefore, that, although bacteria are known which will grow under some of the physical conditions encountered in North Sea reservoirs such as high pressure, cultures able to give oil mobilization are likely to be inactive at reservoir temperatures greater than *ca* 75°C.

CHANNEL PLUGGING

A major hazard of deliberate injection of bacteria followed by growth in the reservoir rock is the danger of channel blocking (although permeability may be restored by acid treatment). The smaller pores may become wholly or partially plugged by cells or cell aggregates. There is evidence, however, that in static conditions bacterial penetration of a reservoir rock would be greater and cause greater permeability reduction in the regions of high initial permeability (Jenneman *et al.*, 1985).

This effect may be turned to the oil operator's advantage in another aspect of EOR, intentional selective blocking (plugging) of high permeability thief zones in reservoirs by encouraging build-up of biomass and polymeric metabolites and so diverting the water flood to sweep fresh areas of the reservoir (Jack *et al.*, 1983; Jenneman *et al.*, 1984). Strains of *Leuconostoc mesenteroides* synthesizing dextran polymer (Ramsey *et al.*, 1985), caused plugging in high permeability (6 Darcy) sand cores (Jack & DiBlasio, 1985). The use of selective plugging systems based upon *L. mesenteroides* is, however, likely to be restricted by inhibition due to high salinity, temperature and pressure.

Microbial plugging systems have been researched for use in the heavy oil reservoirs of Canada and field trials have been proposed in the Lloydminster area on the Alberta/Saskatchewan border (Jack & DiBlasio, 1985). Here problems centre around the difficulties of displacing a viscous treacle-like material with the more mobile water phase in a heterogeneous matrix with zones of high permeability. In these Canadian fields the temperatures (21–33°C) are well below the upper temperature tolerance level of *L. mesenteroides*, 40°C. In deeper fields, use of *Leuconostoc* might still be feasible as, in contrast to *in situ* production of MEOR oil mobilizing agents, the blocking effect of biomass and polymers could be effective much closer to the injector well, within the region cooled by water flooding.

THE CASE AGAINST *IN SITU* MEOR

From an operator's viewpoint (Grist, 1983) MEOR represents an even greater step into the unknown than conventional EOR. As virtually all proposed MEOR systems are microbial adaptations of conventional EOR techniques, normal problems are not avoided but compounded with microbial problems. Not least is the potential for more than one effect operating simultaneously: technically this may appear to be advantageous but, in practice, ill-understood multiple effects make the task of reservoir simulation well-nigh impossible. The lack of control, even compared with chemical EOR, is also of concern: how can MEOR be stopped if it shows unwanted effects in a large reservoir?

Unless the microbial oil displacement process can be quantified in the laboratory in petroleum engineering terms it will not be considered possible to predict the outcome in the reservoir. Large-scale, long-term operations will not be sanctioned unless they are amenable to some form of predictive computer modelling.

The argument that *in situ* MEOR avoids logistic problems of transport to and storage on platforms of EOR chemicals is also invalid as large quantities of feedstock material would undoubtably be required to sustain micro-organisms on the scale needed to generate oil-mobilizing metabolites. In economic terms, simple mass balance calculations may rule out some MEOR options if the mechanism is based on the production of a single metabolite (Grist, 1983).

In several respects application of MEOR techniques would be disliked by the petroleum engineer. Various components of oilfield operations which could be adversely affected by the introduction of microbial cells into the reservoir have been reviewed by Grist (1983). These include the deliberate injection of particulates (cells or cell aggregrates), the danger of bacterial growth in the 'wrong' places, e.g. blocking well casing perforations, in downhole safety valves and in the porous rock matrix and the possible 'souring' of reservoirs.

Some current oilfield production practices would have to be modified for MEOR. Chlorination to effective chlorine residual levels or organic biocide additions would have to cease for the duration of the MEOR treatment leading to the introduction of general contamination from the flood water including sulphate-reducing bacteria. These secondary populations might compete for substrate with the MEOR organisms or biodegrade their oil mobilizing metabolites.

Based on experience from *in situ* MEOR in low temperature, shallow fields, mixed cultures are likely to be more effective and more resilient to changing conditions in the reservoir habitat. This implies a haphazard outcome of reservoir treatment, unacceptable in oilfields with valuable reserves still in place.

CONCLUSIONS

In situ techniques of microbial enhanced oil recovery are unlikely to be adopted for use in deep, hot fields operated from off-shore production platforms. Many factors contribute to the relegation of *in situ* MEOR to small, cool, low risk, low value, on-shore oil fields. Of most importance to the oil-field operator is the lack of data from which to model reservoir EOR simulations and problems of control once an MEOR treatment is under way.

This lack of field data and paucity of thorough research into MEOR process mechanisms and control is an inevitable consequence of the fact that reservoirs suitable for *in situ* MEOR will also be suitable for non-biological methods of tertiary recovery, often with proven efficiencies in laboratory and pilot trials. Because of this, *in situ* MEOR will continue to be used only as a last resort in watered-out fields.

Nevertheless, the amount of oil left behind after secondary recovery methods have been used, represents an attractively large target for extraction and the major oil companies are actively considering conventional EOR processes. In future years, with improved modelling techniques and more experience with various non-microbial EOR methods, time may be ripe for a reconsideration of *in situ* MEOR for suitable fields with significant residual oil levels. At present, selective channel plugging for the extraction of heavy oil from cooler reservoirs appears to be a promising application for microbial activity in enhanced oil recovery; field trial results are, however, required.

ACKNOWLEDGEMENTS

The authors would like to thank the British Petroleum Company plc for permission to publish this paper.

REFERENCES

Baross, J. A. and Deming, J. W. (1983). 'Growth of "black smoker" bacteria at temperatures of at least 250°C', *Nature*, **303**, 423–426.
Britton, L. N. (1984). 'Microbial degradation of aliphatic hydrocarbons', in *Microbial Degradation of Organic Compounds* (ed. D. T. Gibson), pp 89–130, Marcel Dekker, New York.

Brock, T. D. (1985). 'Life at high temperatures', *Science*, **230**, 132–138.

Bubela, B. (1984). 'Effect of biological activity on the movement of fluids through porous rocks and sediments and its application to enhanced oil recovery', *Geomicrobiology J.* **4**, 313–327.

Bubela, B. and Mckay, B. A. (1985). 'Assessment of oil reservoirs for microbiological enhanced oil recovery', in *Microbes and Oil Recovery*, Vol. 1, International Bioresources Journal (eds J. E. Zajic and E. C. Donaldson), pp 99–107, Bioresources Publishing, El Paso, Texas.

Carothers, W. W. and Kharaka, Y. K. (1978). 'Aliphatic acid anions in oilfield waters — implications for origin of natural gas', *Am. Soc. Petroleum Geol. Bull.*, **62**, 2441–2453.

Chakrabarty, A. M. (1985). 'Genetically-manipulated microorganisms and their products in the oil service industries', *Trends Biotechnol.*, **3**, 32–38.

Connan, J. (1986). 'Biodegradation of crude oil in the reservoir', (this volume pp 49–56).

Deming, J. W. (1986). 'The biotechnological future for newly described extremely thermophilic bacteria'. *Microbial Ecol.*, **12**, 111–119.

Donaldson, E. C. and Bennett Clark, J., eds. (1983). *Proceedings of 1982 International Conference on Microbial Enhancement of Oil Recovery*, U.S. Department of Energy, Bartlesville, Oklahoma.

Dow, F. K. and Quigley, T. M. (1986). 'Mechanisms of microbial transport through porous rock', (this volume p 93).

Duvnjak, Z. and Kosaric, N. (1985). 'Production and release of surfactant by *Corynebacterium lepus* in hydrocarbon and glucose media'., *Biotechnol. Letts.*, **7**, 793–796.

Finnerty, W. R. and Singer, M. E. (1983). 'Microbial enhancement of oil recovery', *Bio/Technology*, **1**, 47–54.

Gibson, D. T. ed., (1984). *Microbial degradation of organic compounds*, Marcel Dekker, New York.

Grist, D. M. (1983). 'Microbial enhancement of oil recovery — an operator's view', in 'Biotech '83', pp 463–474. Online Publications, Northwood, UK.

Guerra-Santos, L., Kappeli, O. and Fiechter, A. (1984). '*Pseudomonas aeruginosa*' biosurfactant production in continuous culture with glucose as carbon source', *Appl. Env. Microbiol.*, **48**, 301–305.

Hann, D. and McGillivray, A. (1985). 'Enhanced oil recovery on the UK continental shelf', *Petroleum Review*, December, 36–39.

Herbert, B. N. (1986) 'Reservoir souring', (this volume pp 63–72).

Hitzman, D. O. (1983). 'Petroleum microbiology and the history of its role in enhanced oil recovery', in *Proc. Int. Conf. Microbial Enhanced Oil Recovery 1982* (eds. E. C. Donaldson and J. Bennett Clark), pp 162–218, US Department of Energy, Bartlesville, Oklahoma.

Jack, T. R. and DiBlasio, E. (1985). 'Selective plugging for heavy oil recovery', in *Microbes and Oil Recovery, Vol. I, International Bioresources Journal*, (eds. J. E. Zajic and E. C. Donaldson), pp 205–212, Bioresources Publishing, El Paso, Texas.

Jack, T. R., DiBlasio, E., Thompson, B. G. and Ward, V. (1983). 'Bacterial systems for selective plugging in secondary oil production', *Prepr. — Am. Chem. Soc., Div. Pet. Chem.*, **28**, 773–784.

Jack, T. R., Lee, E. and Mueller, J. (1985) 'Anaerobic gas production from crude oil', in *Microbes and Oil Recovery, Vol. 1, International Bioresources Journal* (eds. J. E. Zajic and E. C. Donaldson) pp 167–180, Bioresources Publications, El Paso, Texas.

Jang, L. K., Findley, J. E. and Yen, T. F. (1983). 'Preliminary investigation on the transport problems of microorganisms in porous media', in *Microbial Enhanced Oil Recovery* (eds. J. E. Zajic, D. G. Cooper, T. R. Jack and N. Kosaric), pp 45–49, PennWell Publishing Co., Tulsa, Oklahoma.

Jang, L. K. and Yen, T. F. (1985). 'A theoretical model of convective diffusion of motile and non-motile bacteria toward solid surfaces', in *Microbes and Oil Recovery, Vol. 1, International Bioresources Journal* (eds. J. E. Zajic and E. C. Donaldson) pp 226–246, Bioresources Publishing, El Paso, Texas.

Janshekar, H. (1985). 'Microbial enhanced oil recovery processes', in *Microbes and Oil Recovery, Vol. 1, International Bioresources Journal* (eds. J. E. Zajic and E. C. Donaldson), pp 54–84, Bioresources Publishing, El Paso, Texas.

Jenneman, G. E., Knapp, R. M., McInerney, M. J., Menzie, D. E. and Revus, D. E. (1984) 'Experimental studies of in-situ microbial enhanced oil recovery', *Soc. Pet. Eng. J.*, **24**, 33–37.

Jenneman, G. E., McInerney, M. J. and Knapp, R. M. (1985). 'Microbial penetration through nutrient-saturated Berea sandstone', *Appl. Env. Microbial.*, **50**, 383–391.

Kosaric, N., Gray, N. C. C. and Cairns, W. L. (1984). 'Microbial emulsifiers and de-emulsifiers', in *Biotechnology* (ed. H. Dellweg), Vol. 3, pp 575–594, Verlag Chemie, Florida.

Lazar, I. (1983a). 'Microbial enhancement of oil recovery in Romania', in *Proc. Int. Conf. Microbial Enhanced Oil Recovery 1982* (eds. E. C. Donaldson and J. Bennett Clark), pp 140–148, US Department of Energy, Bartlesville, Oklahoma.

Lazar, I. (1983b). 'Some characteristics of the bacterial inoculum used for oil release from reservoirs', in *Microbial Enhanced Oil Recovery* (eds. J. E. Zajic, D. G. Cooper, T. R. Jack and N. Kosaric), pp 73–82, PennWell Publishing Co., Tulsa, Oklahoma.

Levi, J. D., Regnier, A. P., Vance, I. and Smith, A. D. (1985). 'MEOR strategy and screening methods for anaerobic oil-mobilising bacteria', in *Microbes and Oil Recovery, Vol. 1, International Bioresources Journal* (eds. J. E. Zajic and E. C. Donaldson), pp 336–344, Bioresources Publishing, El Paso, Texas.

Marquis, R. E., (1976). High-pressure microbial physiology. *Adv. Microbial Physiol.*, **14**, 159–241.

Moses, V. and Springham, D. G. (1982). *Bacteria and the Enhancement of Oil Recovery*, Applied Science Publishers, London.

Parkinson, G. (1985). 'Research brightens future of enhanced oil recovery', *Chem. Engng*, **92** (5), 25–31.

Petzet, G. A. and Williams, R. (1986). 'Operators trim basic EOR research', *Oil & Gas J.*, **84** (6), 41–46.

Ramsey, J. A., Cooper, D. G. and Neufeld, R. J. (1985). 'The effects of reservoir conditions on various bacteria', in *Microbes and Oil Recovery, Vol. 1, International Bioresources Journal* (eds. J. E. Zajic and E. C. Donaldson), pp 108–121, Bioresources Publishing, El Paso, Texas.

Shennan, J. L. and Levi, J. D. (1986). '*In situ*' enhanced oil recovery', in *Biosurfactants and Bioengineering* (eds. N. Kosaric, W. L. Cairns and N. C. C. Gray), to be published, Marcel Dekker, New York.

Sleat, R. and Robinson, J. P. (1984). 'The bacteriology of anaerobic degradation of aromatic compounds', *J. Appl. Bact.*, **57**, 381–394.

Springham, D. G. (1982). 'Bugs to the oil industry's rescue', *New Scientist*, **94**, 408–410.

Springham, D. G. (1984). 'Microbiological methods for the enhancement of oil recovery', in *Biotechnology & Genetic Engineering Reviews, Vol. I*, (ed. G. E. Russell), pp 187–221, Intercept, Newcastle-upon-Tyne.

Stetter, K. O. (1984). Anaerobic life at extremely high temperatures. *Origins Life*, **14**, 809–815.

Sutherland, I. W. (1986). 'Downhole use of biopolymers', (this volume pp 93–104).

Torbati, H. M., Donaldson, E. C., Jenneman, J. E., Knapp, R. M., McInerney, M. J. and Menzie, D. E. (1985). 'Depth of microbial plugging and its effect on pore size distribution in Berea sandstone cores', in *Microbes and Oil Recovery, Vol. 1, International Bioresources Journal* (eds. J. E. Zajic and E. C. Donaldson), pp 213–225, Bioresources Publishing, El Paso, Texas.

Trent, J. D., Chastain, R. A. and Yayanos, A. A. (1984). Possible artefactual basis for apparent growth of bacteria at 250°C. *Nature*, **307**, 737–740.

Vance, I. and Hunt, R. J. (1985). 'Technical note: modification of a French Press for the incubation of anaerobic bacteria at elevated pressures and temperatures', *J. Appl. Bact.*, **58**, 525–528.

Vance, I. and Richmond, B. G. (1985). 'An improved sampling technique for bacterial cultures under high hydrostatic pressure', *J. Micro. Methods*, **4**, 37–43.

White, R. H. (1984). 'Hydrolytic stability of biomolecules at high temperatures and its implication for life at 250°C. *Nature*, **310**, 430–432.

Wiegel, J., Lungdahl, L. G. and Rawson, R. R. (1979). 'Isolation from soil of the extreme thermophile *Clostridium thermohydrosulfuricum*', *J. Bact.*, **139**, 800–810.

Yarbrough, H. F. and Coty, V. F. (1983). 'Microbially enhanced oil recovery in the upper cretaceous Nacatoch formation, Union County, Arkansas', in *Proc. Int. Conf. Microbial Enhanced Oil Recovery 1982* (eds. E. C. Donaldson and J. Bennett Clark), pp 149–153, US Department of Energy, Bartlesville, Oklahoma.

Yayanos, A. A., van Boxtel, R. and Dietz, A. S. (1984). 'High-pressure-temperature gradient instrument: Use for determining the temperature and pressure limits of bacterial growth', *Appl. Env. Micro.*, **48**, 771–776.

Young, L. Y. (1984). 'Anaerobic degradation of aromatic compounds', in *Microbial Degradation of Organic Compounds* (ed. D. T. Gibson) pp 487–524, Marcel Dekker, New York.

Zajic, J. E., Cooper, D. G., Jack, T. R. and Kosaric, N. (eds.) (1983). *Microbial Enhanced Oil Recovery*, PennWell Publishing Co., Tulsa, Oklahoma.

Zajic, J. E. and Donaldson, E. C. (eds) (1985). *Microbes and Oil Recovery, Vol. 1, International Bioresources Journal*, Bioresource Publications, El Paso, Texas.

Zajic, J. E. and Seffens, W. (1984). 'Biosurfactants', *CRC Critical Rev. Biotechnology*, **1**, 87–107.

Zhang, Z. and Qin, T. (1983). 'A survey of research on the application of microbial techniques to the petroleum production in China', in *Proc. Int. Conf. Microbial Enhanced Oil Recovery 1982* (eds. E. C. Donaldson and J. Bennett Clark), pp 135–139, US Department of Energy, Bartlesville, Oklahoma.

Zobell, C. E. and Cobet, A. B. (1962). 'Growth, reproduction and death rates of *Escherichia coli* at increased hydrostatic pressures', *J. Bact.*, **84**, 1228–1236.

Zobell, C. E. and Oppenheimer, C. H. (1950). 'Some effects of hydrostatic pressure on the multiplication and morphology of marine bacteria', *J. Bact.*, **60**, 771–781.

DOWNHOLE USE OF BIOPOLYMERS

Ian W. Sutherland and Christian Kierulf

*Department of Microbiology, University of Edinburgh,
Kings Buildings, West Mains Road, Edinburgh EH9 3JG*

Biopolymers (microbial extracellular polysaccharides) have found a variety of applications in the oil industry. These applications occur at all stages of field development and, in each case, a biopolymer must compete against alternative water-soluble materials in respect of cost, physical properties and effectiveness.

The Nature of Biopolymers

Biopolymers are extracellular polysaccharides synthesized by various micro-organisms. Although a wide variety of such polymers are produced, a relatively small number have been commercially developed; a few of these have in turn found oilfield applications. The polysaccharide which is most familiar to the developer or operator of an oilfield is *xanthan*, the product of the plant pathogenic bacterium *Xanthomonas campestris*. The name is applied to a group of polymers (Table 1) which are essentially substituted celluloses having a pentasaccharide repeat unit, and carrying O-acetyl and pyruvate ketal groups (Figure 1) (Jansson *et al.*, 1975), which may be of

Table 1 Types of 'Xanthan' available

Molar Ratio					
Mannose	Glucose	Glucuronic Acid	Acetate	Pyruvate	
2	2	1	1	0.3	Most commercial products
2	2	1	1	0	Mutant strain developed commercially
2	2	1	1	<0.1	*X. phaseoli* strain
2	2	1	<0.1	0.3	*X. phaseoli* strain
2	2	1	1	~0.7	Commercial product
2	2	1	~2	~0.1	Certain strains and pathovars
2	2	1	0	0	Can be prepared chemically
2	2	1	1	0.3	'Low viscosity' material (patent)
<2	2	<1	0	0	Mutant preparations

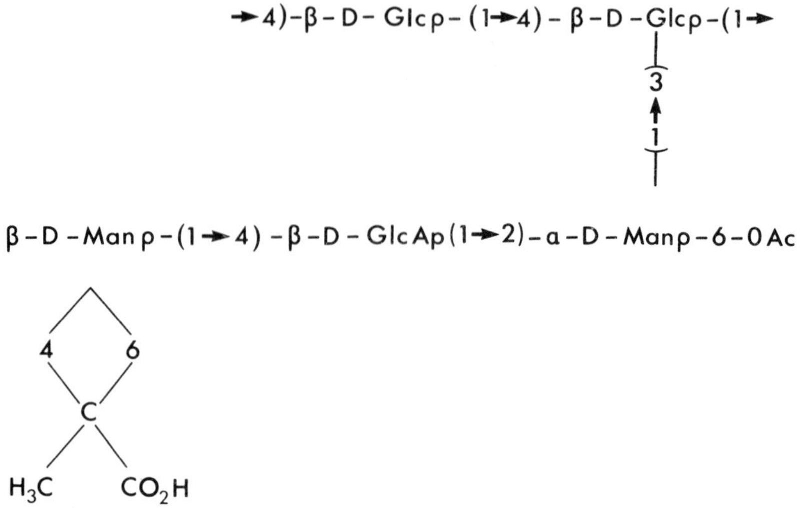

$$\rightarrow 4)-\beta-D-Glc\,p-(1\rightarrow4)-\beta-D-Glc\,p-(1\rightarrow$$

$$\begin{array}{c} \downarrow \\ 3 \\ \uparrow \\ 1 \\ \top \end{array}$$

$$\beta-D-Man\,p-(1\rightarrow4)-\beta-D-GlcAp\,(1\rightarrow2)-a-D-Man\,p-6-0\,Ac$$

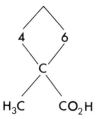

H₃C   CO₂H

FIG 1   The structure of the exopolysaccharide from *X. campestris* (xanthan).

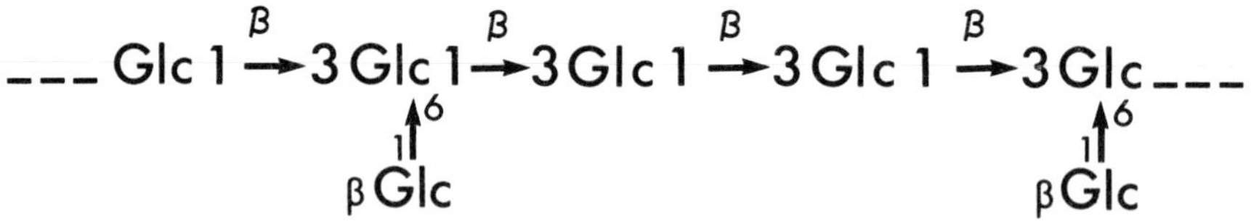

$$\text{---}Glc\,1\overset{\beta}{\rightarrow}3\,Glc\,1\overset{\beta}{\rightarrow}3\,Glc\,1\overset{\beta}{\rightarrow}3\,Glc\,1\overset{\beta}{\rightarrow}3\,Glc\text{---}$$
$$\quad\quad\quad\quad\quad{}_1\!\!\uparrow^6\quad\quad\quad\quad\quad\quad\quad\quad\quad{}_1\!\!\uparrow^6$$
$$\quad\quad\quad\quad\quad\beta\,Glc\quad\quad\quad\quad\quad\quad\quad\quad\quad\beta\,Glc$$

FIG 2   The structure of scleroglucan, the polymer from *Sclerotium* spp. e.g. *S. rolfsii*.

considerable importance in the industrial applications of this polysaccharide. The typical bacterial product is a high molecular weight ($>1m$) polysaccharide (Morris *et al.*, 1983) of relatively uniform size — polydispersity 1.2 (Lambert *et al.*, 1982), yielding highly viscous aqueous solutions. The industrial products used in earlier studies were dry powders presenting considerable problems during rehydration, but liquid concentrates containing approximately 5–6% polysaccharide are now available. There is possibly scope for both biological improvements in the biopolymers and for further technical advances such as this in the method of preparation and delivery. Several other biopolymers are now commercially available, including a neutral fungal homopolymer (scleroglucan) (Figure 2), a polysaccharide from *Sclerotium* species. A bacterial product, S130 is

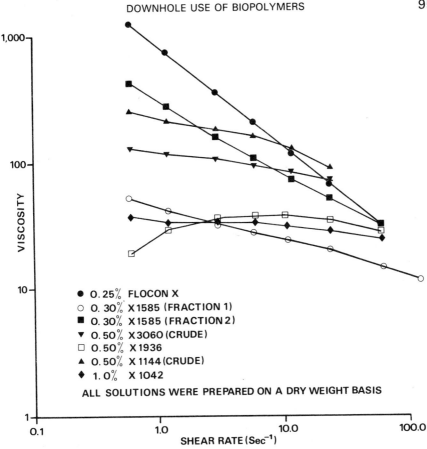

FIG 3   The viscosity of different xanthans over a range of shear rates.

obtained from a strain of *Alcaligenes*. This anionic polysaccharide has recently been shown to be one of a group of closely related chemical structures from a series of biopolymers all of which have interesting and potentially useful physical properties (Jansson *et al.*, 1986). The differing chemical structures of these biopolymers are reflected in their physical properties, particularly as regards solubility and viscosity in aqueous solution, and their stability and compatibility with salts and other chemicals. In oil recovery, as in any other industrial application of polysaccharides, they must compete with other water soluble polymers, which will almost certainly be cheaper to produce. These include polyacrylamides and other synthetic polymers, modified natural products such as hydroxypropyl guar, carboxymethyl- and hydroxyethyl-celluloses and lignosulphonates. All of these although considerably cheaper than biopolymers, are inferior in terms of performance, particularly if xanthan is taken as an example of the latter.

Physical Properties

Aqueous solutions of xanthan are highly viscous pseudoplastic fluids, but marked differences may be found between different preparations (Figure 3). Solutions are easily pumped without showing the shear degradation observed with polyacrylamides. The viscosity is maintained at high salt concentrations and over a wide pH range. This may be due to the structure of the polymer in solution, generally regarded as being an ordered double helical conformation, although various other interpretations have been postulated (e.g. Stokke *et al.*, 1986). This ordered form is lost at high temperatures in the absence of salt and transition from order to disorder is dependent on the ionic environment and on the precise structure of the biopolymer molecule. The presence or absence of acyl groups may exert considerable influence, it being suggested that the ordered structural conformation is stabilized by apolar interaction of the methyl groups of acetyl residues. Pyruvate groups exert a converse effect, strongly destabilizing the ordered conformation due to intramolecular electrostatic repulsion between pyruvate groups (Dentini *et al.*, 1984; Smith *et al.*, 1981). In xanthan solutions, the transition is relatively gradual compared with the effect seen in the melting of biopolymer gels such as that seen from *Enterobacter aerogenes* XM6 (Nisbet *et al.*, 1984). Chemical removal of the acetyl or pyruvate groups can be achieved without destruction of the carbohydrate structure. The products retain their viscosity in solution at least over certain ranges of concentration and shear rate (Bradshaw *et al.*, 1983). It is also possible to add certain chemical groups such as amines to biopolymers and this may also affect the physical properties of the molecule. (Shay and Reiter, 1984). Most of the aminated products were more viscous in brine solution than the native xanthan solutions. The modified polysaccharides also show greater resistance to enzyme systems hydrolysing native xanthan (I. W. Sutherland and J. M. Drury, unpublished results).

Stability

The stability of the biopolymer solution both prior to utilization and *in situ* in the reservoir, are of prime importance.

Biological stability

Under reservoir conditions, if the temperature is high as is the case in the UK offshore area, biological degradation is unlikely. However, during solution preparation at lower temperatures and during injection, the addition of preservatives such as formaldehyde is necessary. All biopolymers are

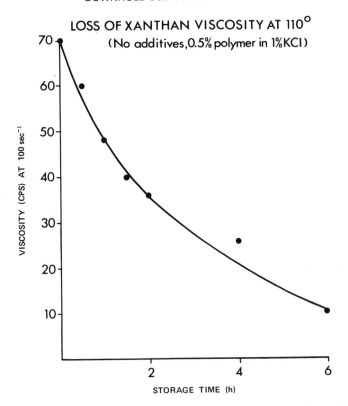

FIG 4   The loss of xanthan viscosity at 110° (in the absence of antoxicants and sacrificial agents).

theoretically biodegradable. In the case of xanthan, several specific enzymes hydrolysing the polymer have been identified and studied (e.g. Lesley, 1961; Sutherland, 1982). This polymer, because of its cellulosic backbone, can also be degraded by some fungal cellulases, but only under conditions where the polysaccharide is in the disordered state (Sutherland, 1984). Thus such enzymes are not likely to cause significant degradation of xanthan solutions at ambient temperatures, particularly when salts are likely to be present and assist in promoting the ordered structure.

   The presence of antimicrobial agents is also necessary to ensure that no microbial growth occurs due to the presence of small quantities of bacterial cells, cell constituents such as proteins, DNA and RNA, and unutilized components of the bacterial growth medium. If micro-organisms were to grow on such components, even without producing biopolymer-degrading enzymes, the microbial cells would cause appreciable problems through introduction of particulate material and reduced injectability.

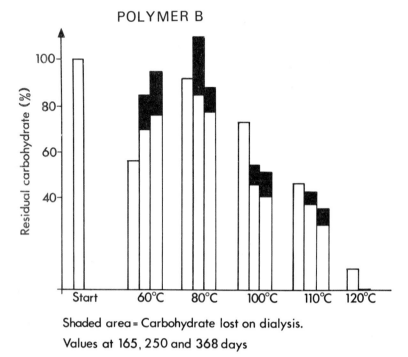

POLYMER B

Shaded area = Carbohydrate lost on dialysis.

Values at 165, 250 and 368 days

FIG 5   The stability of a xanthan preparation as measured by carbohydrate analysis. The diagram indicates the amount of carbohydrate recoverable after exposure to various temperatures for prolonged periods in the absence of oxygen. The shaded area indicates the portion of low molecular weight material lost on dialysis.

Chemical stability

The chemical stability of any biopolymer is a function of its chemical structure. Few biopolymers have yet proved to be as stable as xanthan, but all are susceptible to both hydrolytic and oxidative degradation. At pH values near neutrality, there is less likelihood of hydrolytic depolymerization, but oxidative degradation is a major problem under the conditions where biopolymers are likely to be used. The problem is accentuated by the presence of iron, which catalyses free radical reactions and thus stimulates degradation (Parsons *et al.*, 1985). The susceptibility of microbial polysaccharides to depolymerization is reflected by three aspects: i) destruction of carbohydrate; ii) release of low molecular weight material due to breakage of the glycosidic bonds of the polymers; and iii) loss of viscosity. In xanthan, the most labile linkages are likely to be the α- and β-mannosidic bonds at the termini of the side-chains. If main-chain cleavage occurs to even a very small number of the 1,4 β-glucosidic linkages, rapid and irreversible loss of viscosity occurs (Sutherland, 1984).

Loss of viscosity in xanthan solutions is rapid at high temperatures if no attempt is made to exclude oxygen or 'protect' the biopolymer with sacrificial agents and antoxidants (Figure 4). If such precautions are taken however, the polymer solution is relatively stable at least up to temperatures of 90°C. The 'half-life' of one xanthan preparation as measured by total carbohydrate estimations, was 1.51 years and 0.74 years at temperatures of 80°C and 100°C respectively. Recent results in our laboratory suggest that, with improved polymer preparations and better protection, together with more accurate analytical procedures, such values may be underestimates (Figure 5). One must however distinguish between the different aspects of stability. Results for a laboratory preparation are however encouraging, showing a slight but relatively gradual destruction of carbohydrate over 368 days in the temperature range 80–90°C. Viscosity was also well maintained at 90°C over this time period (Figure 6). The final value at 80°C was probably an artifact due to problems encountered in accurately maintaining this temperature.

Downhole Applications

Three major applications of biopolymers for downhole use exist. The major use to date has been at the level of exploration and field development in drilling, completion and gravel packing. In such applications, use is made of the biopolymers primarily for viscosity control and suspending capacity. Patents using the polysaccharide xanthan as a component of drilling muds date from 1966. As such it has become an accepted chemical for oilfield use. As deeper wells are drilled, biopolymers superior to xanthan in their

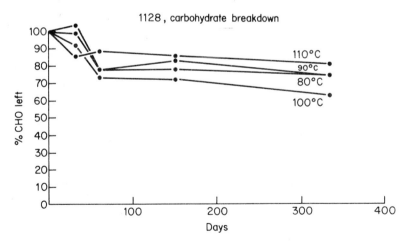

FIG 6(a) The stability of a laboratory xanthan preparation as measured by carbohydrate analysis.

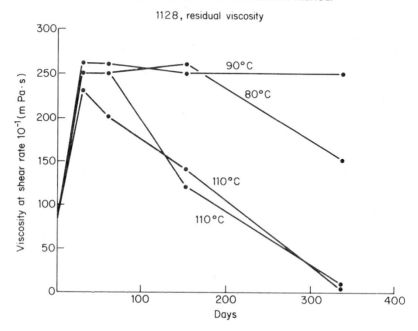

FIG 6(b) The stability of the same material as measured by viscometry.

ability to withstand degradation at high temperatures may be required. One such S130 polymer, has recently been reported as having superior suspending capacity to hydroxyethyl cellulose or hydroxypropyl guar and viscosity retention at temperatures up to 150°C (Sandford et al., 1984), although it is not compatible with high density brines. Solutions are also claimed to be almost unaffected over the pH range 2–12. It has been suggested that S130 polymer solutions would be suitable for suspension of cuttings, bridging particles, weighting material, fracturing sand or gravel for packing (Baird et al., 1983). Solutions of S130 are highly pseudoplastic and could be ideal for use as a constituent of low solids drilling muds.

Stability of biopolymers for operations such as these, and compatibility with other oilfield chemicals are important, but the duration of the operation may well be short compared to EOR. Thus polysaccharides which are stable for short time periods may fulfil the necessary requirements. There may also be less of a need for high injectivity as the fluids will not normally progress from the well structure into the reservoir rocks.

The greatest potential application for biopolymers is in Enhanced Oil Recovery. This would require much greater quantities of the polymers than other operations. The requirement for high stability and good injectivity can be met by xanthan, especially if any residual particulate material is removed. This can readily be accomplished by enzymic treatment (Figure 7) (Griffith

FLOCON TREATMENT WITH PROTEASE

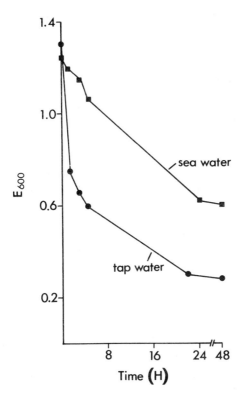

FIG 7   The use of proteases to remove particulate material from xanthan solutions. The effect of proteases on solutions can be followed through measuring the turbidity of the solution.

et al., 1981; Symes and Smith, 1983). The high shear stability of biopolymers such as xanthan ensures that viscosity is not reduced after solutions have been subjected to the high shear conditions existing during mixing and injection. The pseudoplasticity of the solutions renders them relatively easy to pump. As stability for prolonged periods *in situ* is of pre-eminent importance, a number of studies have been made to test this, as already indicated. While xanthan has been the polysaccharide most widely tested, others such as ENORFLO-S, a product from a *Pseudomonas* sp., have also been evaluated (Ash *et al.*, 1983). The advantage of xanthan over polyacrylamides was demonstrated in alkaline flood experiments by Auerbach (1984), but it has to be remembered that both biopolymers and synthetic compounds have to be adequately protected from oxygen to ensure stability of the polymer solutions in reservoirs (Knight, 1973).

As EOR is likely to involve the simultaneous injection of other chemicals, the compatibility of biopolymers with reservoir fluids containing high concentrations (200–100, 000 ppm.) of monovalent and divalent cations is important. Equally, xanthan solutions can be combined with alkaline agents such as sodium hydroxide or sodium orthosilicate used in alkali floods, or (for shorter time periods) with relatively strong acids such as 15 N HC1. Not all surfactants are compatible with biopolymers and careful choice of non-reacting compounds may be necessary. Under some conditions, xanthan solutions may influence mobility control through interaction of the polysaccharide molecules with the surfaces of the rock pores (Duda *et al.*, 1981). Mobility reduction by xanthan was less than that for polyacrylamides and the permeability reduction was considered to be lower, but during the flow of the polymer solution a layer of adsorbed biopolymer molecules and molecules trapped in pore constrictions can reduce the effective permeability.

The third downhole application of biopolymers is for profile modification. The lack of uniformity in the permeability of reservoir rocks leads to considerable losses of EOR chemicals and fluids into 'thief zones'. Biopolymer solutions can be converted *in situ* into cross-linked gels through the addition of trivalent chromium ions, thus leading to increased sweep efficiency of the water floods. Following injection of premixed xanthan and cross-linking agent, the reaction is relatively rapid. Because of shear-thinning during injection, the polysaccharides can readily be introduced into the high permeability zones. If necessary, the gels can eventually be degraded with an oxidizing agent such as hypochlorite. As with EOR applications, biopolymer gels may be resident under reservoir conditions for relatively, long periods and laboratory tests have shown that the gels remain stable for up to 10 months at ˜90°C (Chang *et al.*, 1985).

Whether other biopolymers will be as suitable as xanthan for *in situ* gelation remains to be determined. It is however clear that xanthan is an extremely versatile biopolymer which can be produced on a large scale, near the oilfield if necessary or alternatively readily transported as a concentrated fluid of consistent quality. Any other microbial polysaccharide to replace xanthan would have to possess superior physical attributes and higher cost effectiveness. Future developments in the application of biopolymers will probably be to a large extent dependent on the financial return for oil field developments; it is only to be hoped that the current downward trend in oil prices will not prove too inhibitory to further work on biopolymers for oil industry applications.

REFERENCES

Ash, S. G., Clarke-Sturman, A. J., Calvert, R. and Nisbet, T. M. (1983). 'Chemical stability of biopolymer solutions', *SPE Paper* 12085.

Auerbach, M. H. (1984). 'Mobility control polymers in alkalne flood EOR', *Amer. Chem. Soc. Symp. on Chemistry of E.O.R.*

Baird, J. K., Sandford, P. A. and Cottrell, I. W. (1983). 'Industrial applications of some new microbial polysaccharides', *Biotechnol.*, **1**, 778–783.

Bradshaw, I. J., Nisbet, B. A., Kerr, M. H. and Sutherland, I. W. (1983). 'Modified xanthan — its preparation and viscosity'. *Carbohydr. Polymers*, **3**, 23–28.

Chang, P. W., Goldman, I. M. and Stingley, K. J. (1985). 'Laboratory studies and field evaluation of a new gelant for high-temperature profile modification'. *SPE Paper* 14235.

Dentini, M., Crescenzi, V. and Blasi, D. (1984). 'Conformational properties of xanthan derivatives in dilute aqueous solution', *Intern. J. Biol. Macromol.*, **6**, 93–98.

Duda, J. L., Klaus, E. E. and Fan, S. K. (1981). 'Influence of polymer molecule/wall interactions on mobility control', *Soc. Petrol. Eng. J.*, pp 613–622.

Griffith, W. L., Compere, A. L. and Holleman, J. W. (1980). UK Patent 2065688A.

Jansson, P. -E., Kenne, L. and Lindberg, B. (1975). 'Structure of the extracellular polysaccharide from *Xanthomonas campestris*', *Carbohydr. Res.*, **45**, 275–282.

Jansson, P. -E., Lindberg, B., Widmalm, G. and Sandford, P. A. (1986). 'Structural studies of an extracellular polysaccharide S130 elaborated by *Alcaligenes* ATCC 3155', *Carbohydr. Res.*, (in press).

Knight, B. L. (1973). 'Reservoir stability of polymer solutions', *J. Petrol. Technol.*, pp 618–621.

Lambert, F., Milas, M. and Rinaudo, M. (1982). 'Gel permeation chromatography of the xanthan gum using a light scattering detector', *Polymer Bull.*, **7**, 185–189.

Lesley, S. M. (1961). 'Degradation of the polysaccharide of *Xanthomonas phaseoli* by an extracellular bacterial enzyme', *Canad. J. Microbiol.*, **7**, 815–825.

Morris, V. J., Franklin, D. and I'Anson, K. (1983). 'Rheology and microstructure of dispersions and solutions of the microbial polysaccharide from *Xanthomonas campestris* (Xanthan gum)', *Carbohydr. Res.*, **121**, 13–20.

Nisbet, B. A., Sutherland, I. W., Bradshaw, I. J., Kerr, M. H., Morris, E. R. and Shepperson, W. A. (1984). 'XM6: A new gel-forming bacterial polysaccharide', *Carbohydr. Polymers.*, **4**, 377–394.

Parsons, B. J., Phillips, G. O., Thomas B., Wedlock, D. J. and Clarke-Sturman, A. J. (1985). 'Depolymerization of xanthan by iron-catalysed free radical reactions', *Intern. J. Biol. Macromol.*, **7**, 187–192.

Sandford, P. A., Cottrell, I. W. and Pettit, D. J. (1984). 'Microbial polysaccharides: new products and their commercial applications', *Pure Appl. Chem.*, **56**, 879–892.

Shay, L. K. and Reiter, S. E. (1984). US Patent 4485020.

Smith, I. H., Symes, K. C., Lawson, C. J. and Morris, E. R. (1981). 'Influence of the pyruvate content of xanthan on macromolecular association in solution', *Int. J. Biol. Macromol.*, **3**, 129–134.

Stokke, B. T., Elgsaeter, A. and Smidsrod O. (1986). 'Flexibility of single and double-stranded xanthan. An electron microscopic study', (in press).

Sutherland, I. W. (1982). 'An enzyme system hydrolysing the polysaccharides of *Xanthomonas* species', *J. Appl. Bact.*, **53**, 385–393.

Sutherland, I. W. (1984). 'Hydrolysis of unordered xanthan in solution by fungal cellulases', *Carbohydr. Res.*, **131**, 93–104.

Symes, K. C. and Smith, I. H. (1983). UK Patent 2111520A.

ENVIRONMENTAL ASPECTS OF OFFSHORE OPERATIONS

H. J. Somerville

Shell UK Exploration and Production, Aberdeen, Scotland

INTRODUCTION

Within recent years there has been a resurgence of interest in the overall environmental health of the North Sea. This resurgence was marked by the first ministerial North Sea Conference held in 1984, with a second such conference to be hosted by the United Kingdom in 1987.

This interest in the North Sea reflects a trend to examine environmental issues in regional terms related to identifiable mixing zones, particularly for aqueous discharges. Thus, international conventions have been established covering the North Sea and adjacent waters (for example, 1974 Paris Convention); the Baltic (1974 Helsinki Convention), and the Mediterranean (1976 Barcelona Convention). These conventions are of relatively long standing indicating that interest in the environmental health of the oceans is not new.

Within the present symposium it would seem appropriate to review the inputs to the marine environment from the offshore industry in the context of the North Sea as a receiving body and with particular regard to the interactions with micro-organisms. It is inevitable in making such an overview presentation that approximations must be made including extrapolations from very limited information in some cases. Broad interpretations are made not with a particular claim to accuracy but rather to stimulate development of understanding of the fate of discharges to the North Sea, and to promote reasoned identification of any real or potential environmental problems.

REGULATORY BACKGROUND

The North Sea is governed by a range of international conventions affecting differing environmental considerations. In general, these are implemented by specific national legislation, some of which may have been in effect before ratification of the international agreement. Some of the conventions and the means of their implementation in the UK are indicated in Table 1.

Thus it can be seen that there is in existence a comprehensive framework of international legislation which has been implemented in the United Kingdom.

Table 1　International conventions covering environmental aspects of the North Sea

International	Implementation in UK	Area covered
Geneva Conventions 1958	Continental Shelf Act 1964	Pollution prevention control of offshore activities
Oslo Convention 1972	Food & Environment Act 1985	Non-routine discharges
Marpol 1973 (Protocol 1978)	Merchant Shipping Regulations 1983 (Merchant Shipping Act 1979)	'Machinery drainage'
Paris Convention 1974	Prevention of Oil Pollution Act 1971 – 1983 Regulations	Process oily water, oiled drilling cuttings
	Petroleum & Submarine Pipelines Act	Pipeline discharges
	Control of Pollution Act 1984	Nearshore aqueous discharges
Bonn Agreement 1969	Continental Shelf Notice 7 1984	Oil spill contingencies
Convention on Civil Liability 1969	Merchant Shipping (Oil Pollution Act)	Oil spill insurance
'Fund' Convention	OPOL – Voluntary Agreement	Oil spill insurance

Within the Paris Convention, which covers inputs from land-based sources, including production platforms, new regulatory guidelines are generated by discussion at one or more technical levels before any agreement on standards for implementation by individual governments. This ensures a sound technical dialogue addressing the whole of the North Sea.

At present the approach to regulatory control varies for different emissions. For example, the oil level in discharges of production water from North Sea platforms is subject to a standard of 40 ppm with possibility of limited excursions. By contrast, discharges of oil on cuttings from drilling are not subject to any fixed limit although rigorous environmental surveys are carried out and analysis of oil discharges must be reported. This difference arises because for water discharges there is a value which is generally achievable by practicable technology, on the other hand there are so many variables contributing to the actual level of oil on cuttings that any fixed values would be impracticable to impose. Thus, both the emission standard and environmental quality approach to regulatory standards are in current use for the offshore oil and gas industry. The former has been validated by environmental monitoring of water quality around offshore installations.

INPUTS FROM OFFSHORE OIL AND GAS ACTIVITIES

Various aspects of inputs from North Sea Exploration and Production Operations have been previously reviewed (Wilkinson, 1982; Bedborough and Blackman, 1986; Institute of Offshore Engineering, 1985; Somerville, 1984). Current estimates (1984) of some of the major inputs are shown in Tables 2–5. Where Shell Expro figures have been used as a basis these have been multiplied by 4 to give industry totals; this factor has been developed by comparison of Shell Expro and UK totals where both sets of data are available.

Produced Water

Produced water is the name commonly given to water discharged as a result of oil production processes. More correctly it should be defined as water produced from the reservoir which will contain both water from the formation (formation water) and, in relevant fields and wells, injected sea water which has broken through to producing wells. In some production systems with storage cells involving sea-water displacement some displacement water will be included in the final overboard discharge.

Oil in Produced Water

The total of oil in water discharges to the North Sea is of the order of 1,500 tonnes per year (Department of Energy, 1985). The measurement of oil

Table 2   Estimates of some major discharges to the North Sea from UK Exploration and Production Operations (Based on 1984 data)

Input	Quantity tonnes.y^{-1}	Oxygen Demand[a] tonnes.y^{-1}	Biomass[a] tonnes.y^{-1}
Produced water			
Oil (Dept.			
of Energy, 1985)	1430	5000	400–600
Organic compounds[b]	50000	50000	6000–10000
Hydrogen sulphide[b]	1000–1500	2000–3000	—
Ammonia[b]	1000–1500	0	—
Oil Spills (Dept.			
of Energy, 1985)	130	600	25–50
Sewage and domestic[b]	500–1000	500–1000	60–120
Chemicals			
Production (Bedborough			
& Blackman, 1986)	3200	3000–5000	400–800
Drilling (Bedborough			
& Blackman 1986)	5300	5000–7000	600–1000
Pipelines[b]	100	100–150	15–25
Drilling cuttings			
Oil (Dept.			
of Energy, 1985)	20000	15000–20000^2	2000–3000
Others			
Flare drop-out,[b]			
shipping etc.	200–1500	500–3000	100–300

[a]Calculated on basis of 70–90% of organic material being mineralized by microbial action except for oil on drilling cuttings for which only partial degradation (25–30%) is assumed in any one year. The remaining carbon is assumed to be converted to biomass.
[b]Shell Expro estimate extrapolated to UK total.

Table 3   Products discharged as a result of offshore production operations (Bedborough and Blackman, 1986)

	UK total tonnes.y^{-1}	
	1982	1984
Scale inhibitors	950	1200
Corrosion inhibitors	500	220
Biocides	180	500
Demulsifiers	30	600
Oxygen scavengers	30	120
Gas treatment	—	520

Table 4   Products discharged as a result of offshore drilling operations (Bedborough and Blackman, 1986)

	UK total tonnes.y^{-1}	
	1982	1984
Weighting agents and inorganic gelling products	112400	111500
Inorganic chemicals	12560	11000
Polymeric viscosifiers	4280	3680
Lignosulphonates, lignites etc.	700	630
Minor additives		
surfactants, detergents	250	200
de-foamers	65	400
biocides	150	20
corrosion inhibitors	150	20
drilling lubricants	190	200
oxygen scavengers	160	65
dispersants	30	95
pipe release agents	25	15
scale inhibitors	10	10

Table 5   Assessment of litter items on beaches close to an exploratory drilling operation in the Moray Firth

Item Category	1983	1984	Possible Recounts	1985	Possible Recounts	New Items
Fishing Industry	205	146	146	191	146	45
Plastics	239	330	239	317	317	0
Metals	25	82	25	88	82	6
Wood	311	323	311	327	323	3
Glass	10	6	6	5	5	0
Rubber	11	18	11	13	13	0
Paper/Board	9	2	2	20	2	18
Miscellaneous	0	37	0	19	19	0
Total	810	944	740	980	907	73
Percentage Recounts			78.4		92.6	

Litter surveys were carried out at transects on five beaches in the Wick area in October 1983, June 1984 and September 1985 and items identified according to the categories listed above. An exploratory drilling operation was carried out some five miles offshore between December 1983, and April 1984. The majority of the 1985 paper/board items were waxed milk containers at one of the sites. Only one item which could be directly attributed to the offshore Oil and Gas industry was found, in September 1985.

includes hydrocarbons and a small amount of a few hydrophobic compounds derived from crude oil. Although not all components will be equally susceptible to natural degradation processes the chemical compounds in crude oil, particularly in a dispersed form, are generally degradable (Higgins and Gilbert, 1978; Watkinson, 1978). This is particularly true of North Sea crude oil which is light and low in sulphur. Assuming that biodegradation is the major route of degradation there will be an oxygen demand of some 5,000 tonnes per annum and formation of some 400–600 tonnes of biomass, largely as dispersed micro-organisms. It is possible that very small amounts of some of the more recalcitrant components will reach the sea-bed.

A single large production platform might discharge some 150 tonnes of oil per year. In general, the amount of oil discharged with produced water will grow in line with water discharges (Bedborough and Blackman, 1986), but is unlikely to increase by more than a factor of 2–3.

Other Organic Components of Produced Water

Produced water contains substantial amounts of hydrophilic compounds that are derived from the water in the formation and not from the crude oil (Somerville, 1984). An estimated 50,000 tonnes of these compounds, almost totally made up of acetic, propionic and butyric acids, is presently discharged annually in the U.K. sector. These organic acids are readily biodegradable and an estimated 50,000 tonnes of oxygen will be required with formation of some 6,000–10,000 tonnes of biomass.

For a large production platform the amount of organic material discharged might be 3,000–4,000 tonnes per year. The total discharged will increase in the next few years but not in direct relationship with water discharges which will include an increasing proportion of injected sea-water.

Inorganic Components of Produced Water

Produced water contains low levels of many metals (Goodman and Troake, 1984). However, accurate analysis of such water is extremely difficult as formation water and produced injected water will have travelled through many thousands of feet of metal tubing before sampling. There is some variation on a well to well and field to field basis but at this time there is no reason to believe that there are appreciable discharges of metals of concern.

An important inorganic component of produced water is ammoniacal nitrogen which is present in formation water at 20–30 mg.l^{-1}. Thus a total of 1,000–1,500 tonnes of ammonia is discharged per year. As available nitrogen is generally held to be limiting to carbon turnover in the North Sea, no oxygen demand will result. Indeed the ammonium salts in produced water will promote growth and metabolism of marine organisms including the micro-organisms involved in degradation of organic compounds discharged. Again for a large production platform the amount discharged might be 80–120 tonnes per year. Ammoniacal nitrogen discharges may increase in line with the increase in organic components.

Another inorganic compound discharged at some production locations is hydrogen sulphide. Oil and gas from some North Sea reservoirs contain small but significant levels of H_2S. Also, in those installations with storage cells, an environment is created where sulphate from displacement sea-water can be reduced by sulphate reducing bacteria using organic acids in the formation water, which is also directed to the storage cells, as direct or indirect carbon and energy source. An alternative source of carbon and energy for sulphate reducing bacteria in storage cells is the crude oil. It is possible to envisage some partial degradation of crude oil to compounds utilizable by sulphate-reducing bacteria. This is feasible because of the import of seawater to balance import and export, making available oxygen required for initial microbial metabolism of hydrocarbons.

Obviously, levels of H_2S in produced water from these platforms will vary considerably, depending on the extent of mixing, residence time, temperature etc. Measurements by Shell Expro suggest that a total of 1000–1500 tonnes of H_2S may be discharged per year with an associated demand for some 2000–3000 tonnes oxygen. A typical large concrete platform might discharge 200 tonnes of H_2S per year with a demand for about 400 tonnes of oxygen. No allowance has been made in these figures for any H_2S originating in the reservoir.

Sewage and Domestic Waste

Assuming a total working population on platforms and drilling rigs offshore at any given time of about twelve thousand, the sewage inputs can be calculated as approximately 500 tonnes dry weight per year and kitchen waste at about the same. Combined, this will exert a demand of some 500–1000 tonnes oxygen per year. For a typical large production platform some 10–15 tonnes waste could be discharged per year. These inputs depend almost directly on working population.

Chemicals and Drilling Discharges

Selection of process chemicals for use offshore involves a voluntary notification scheme administered by the Department of Energy. The quantity of any particular chemical that can be used per year at any one installation is determined by comparitive rating according to standard toxicity tests.

A detailed examination of the uses of chemicals offshore (Vase, 1983) is not within the scope of this paper. A large part of the total quantity is taken up by a few chemicals, for example within Shell Expro, 70% of the weight used of production chemicals consists of monoethylene glycol, methanol and triethylene glycol which are largely carried through the process in product streams. Of the drilling chemicals the vast bulk consists of inert components; many of the other chemicals are essentially inert or related to natural products such as sodium bicarbonate, soda ash, guargum, lignosulphonates, lime, caustic soda etc.

It is difficult to estimate accurately the quantities of chemicals discharged to the Sea as most process chemicals will partition largely into the oil phase, and, in drilling, there may be substantial losses to the reservoir. Further, many chemicals are supplied in formulations in which water or other innocuous components make up a large part of the mass. Total overall carbon balance studies on produced water in Shell Expro have suggested that offshore production process chemicals are possibly present at about 5 mg.l^{-1} in produced water although no accurate determination of individual chemicals has yet been carried out. This would suggest a total for the UK of some 500 tonnes.y^{-1} of organic process chemicals. This total is somewhat lower than that calculated by the Department of Energy (Table 3). As improved analyses become available it will become possible to generate more accurate figures.

Approximately 5,000 tonnes of organic drilling chemicals were discharged in 1984 (Table 4). Much of this will be closely associated with drilling cuttings which fall to the bottom and which will exert oxygen demand over a more extended timescale than for dispersed discharges.

In addition to the above discharges some 20,000 tonnes of oil on drilling cuttings were discharged in the UK sector in 1984. For Shell Expro the total was approximately 7,000 tonnes in 1984 and more than 75% of the total was 'low toxicity' base oil. This type of oil is now in use for all drilling in the UK. To calculate overall oxygen demand it has been assumed that some 25% of the discharged oil on cuttings will be available for oxidation in the year of discharge. The discharge of chemicals from pipelines is largely associated with hydrotesting and other activities associated with commissioning; consequently there will be a wide variation year to year depending on construction activity. Shell Expro discharges in 1984 were associated with the construction of the Fulmar gas line. In this case it is unlikely that the annual industry total is four times the Shell Expro total, and a factor of two has been used.

The total oxygen demand for degradation of chemicals and drilling base oil is likely to be 20,000–35,000 tonnes per annum with formation of some 3,000–5,000 tonnes of biomass, again assuming biodegradation as the major route and allowing for partial availability of drilling base oil. It is unlikely that this total will increase markedly, even with the possible introduction of new process approaches such as chemical and surfactant flooding which will require careful control measures to minimize discharges.

Debris and Litter

Dumping from marine structures without a licence is prohibited (Table 1). There is also a regulatory requirement for any operator to make every reasonable effort to recover debris that is dropped incidentally (Department of Energy, 1981) and to clean the sea-bed within 70 metres of abandoned well heads (Department of Energy, 1979). Within Shell Expro, emphasis is laid on general good practice both on installations and on supply boats.

An industry fund has been established through UKOOA for dealing with non-Company specific claims from the fishing industry in Scottish Waters; this is administered by the Scottish Fishing Federation without any direct industry input. For example, there were 67 claims for debris damage in the UK in 1982.

Within Shell Expro debris is periodically surveyed and recovered within the 500 metre exclusion zone around the platforms. Each report of debris is investigated and if related to Shell Expro activities, recovery is effected. Additionally Shell Expro each year trawls an area 1 km square around three to four abandoned well sites including any in environmentally sensitive areas, for example in the Moray Firth.

Growing emphasis is being laid on good housekeeping and on accurate reporting of debris lost. It is to be expected that the industry record will continue to improve.

Litter from offshore installations is compacted and returned to shore for disposal. The contribution of offshore oil and gas operations to beached marine litter is low. This has been confirmed by recent surveys of litter on several beaches in Caithness before and after a nearshore exploratory operation (Table 5).

Table 6 Total combined sector North Sea oil inputs from offshore operations by source. 1982/1983 ('000 tonnes) (Institute of Offshore Engineering, 1985)

Production and displacement water	3.0	increasing annually
Oil on cuttings	20.0[a]	reflects annual usage of oil-based muds and drilling activity
Drainage	0.1	relatively constant
Atmospheric inputs from flaring	0.2–1.0	approximate figure, decreasing
Spills	0.4–2.0	highly variable from one year to the next, range makes no allowance for major incidents such as prolonged blowout
Total North Sea	23.7–26.1	

[a]no input from Netherlands Sector drilling activity assumed.

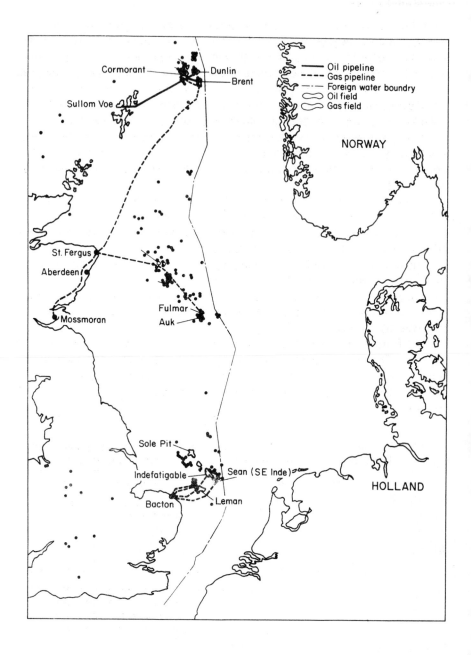

FIG 1   Shell Expro operated wells to end 1985.

Other inputs

Atmospheric inputs to North Sea waters from offshore operations have been estimated (Table 6). Despite the inadequacy of procedures for estimating most of the sources, it is clear that atmospheric inputs are small compared to land-based sources and that any effect on the marine environment is minimal.

Other discharges to water include discharges arising from sacrificial anodes used in corrosion protection, with inputs of some 300 tonnes of zinc and 1000 tonnes of aluminium per year. Discharges of sewage, domestic waste and of 'machinery bilge' drainage water (as covered by the MARPOL convention) from ships, mobile units and platforms are small in total quantity, probably of the order of 100–200 tonnes.y^{-1}. None of these discharges will impact significantly on the environment.

Capacity of the Environment

Quantitatively a total demand of some 200–300 tonnes of oxygen per day is exerted, assuming biodegradation of most components. For a major installation this translates to an oxygen transfer rate of 300–400g. $day^{-1}.m^{-2}$ for an area of radius 100 m around a major platform. This demand is well within calculated transfer rates (J. Bryers and G. Hamer, personal communication) and the absence of oxygen depletion has been confirmed by direct measurement in the immediate vicinity of major offshore discharges.

The total volume of the water body in the UK sector of the North Sea can be roughly estimated at 10^{13} to $10^{14} m^3$ with a much larger water volume in the deeper waters of the North Sea (Figure 1). Some idea of the capacity of the North Sea can be illustrated by assuming a cumulative total of 200,000 tonnes of oil discharged as a result of offshore exploration and production and that the hydrocarbons were totally recalcitrant and distributed evenly in the water. The concentration arising from exploration production operations would be less than 1 in 10^8 by dilution in 'UK water' only with corresponding figures for the other discharges described above. The organic compounds in all these discharges are generally biodegradable with the result that residual concentrations, even in the vicinity of the platform, will be insignificant for those discharges that are dispersed in the water column. These generalizations do not take into account turnover for the North Sea which has, for example, been calculated at 530 days (Prandle, 1984), giving replenishment of water over time scales of significance with respect to natural degradative processes.

The localized effect of oil from drilling discharges has been described previously (Davies et al., 1984). These localized effects arise largely as a

Table 7   Some sediment heavy metals Brent A — 1982

Distance metres	Ba	Zn	Cu	Pb	Fe
			μg/g sediment		
200	2280	250	16.5	60	6000
300	2080	45	5	29	4200
500	1380	30	4	19	4500
1000	300	13	3	16	3960
2000	510	15	3	16	4290
North Sea range	20–1000	12–16	2.5–3.2	8–13	2000–4000

Table 8   Sediment microbial levels Brent A — 1982

Distance (South) from platform	Bacteria levels (No./gramme)		Hydrocarbon levels
metres	aerobes	SRB	ppm
200	10^5–10^6	10^4	1992
300	10^5–10^6	10^2	33
500	10^5–10^6	10^2	22
1000	10^5–10^6	10	15
2000	10^5–10^6	10	19

result of accumulation of inorganic residues around installations with associated organic compounds including drilling base oil as the major component. Metal concentrations in sediments around production platforms also fall quickly to background levels (Table 7). A summary of the environmental implications of the discharge of drilling cuttings has been agreed by the Paris Commission (Paris Commission, 1984). The absolute requirement for molecular oxygen for biodegradation of hydrocarbons will inevitably mean that regeneration of such environments will involve a number of interactive processes in which different microbial and macrobial species including sulphate reducing bacteria (Table 8) and bioperturbative macrobenthic species will play a role. A complex community with gradients of oxygen, partial metabolites and other nutrients and physicochemical parameters will develop. In the long term, there is no reason to believe that unacceptable residues will remain. Already there are encouraging indications that sea-bed regeneration takes place over periods of a few years.

As a result of overall biodegradation processes some 10,000–16,000 tonnes of biomass will be added annually largely in the form of microbial cells. This is equivalent to that fixed annually, as a result of photosynthetic activity, over an area of some 10 kilometres squares, i.e. less than 0.05% of the natural productivity of the UK sector. This will be in a form that in general composition should be indistinguishable from normal primary biomass and thus should be readily assimilated into the food chain.

The ability of micro-organisms to degrade organic compounds has been widely exploited through the major biotechnology of effluent treatment. In aerobic biotreatment typical substrate-stimulated oxygen uptake rates (Somerville, 1985) are in the range $10-200 \, mg.h^{-1}.g \, biomass^{-1}$. This would allow, for a single stage stirred reaction, treatment of some $75-1500 \, g$ organic acids $h^{-1}.m^{-3}$ of biotreating capacity. These figures would suggest that for a large production platform the volume of a biotreater would be in excess of $250 \, m^3$, at least with present technology. The net effect would be to transfer water to the installation with little net environment benefit as recalcitrant compounds and biomass would still be discharged. Thus there appears to be neither any basic reason to biotreat water offshore or any potential gain in environmental terms, although there might be selective application to small concentrated streams in sensitive areas. Any biotreatment process for drilling cuttings would also imply a considerable weight penalty, largely as a result of retention of water on the installation.

EFFECTS OF MICRO-ORGANISMS ON INSTALLATIONS AND OPERATIONS

No overview of the environmental interactions of the offshore oil and gas industry would be complete without consideration of the potential effects of micro-organisms on the operations. Some of these interactions are discussed in detail elsewhere in the present conference. The main areas of Shell Expro experience are discussed briefly below.

Hydrogen Sulphide Generation

Levels of $50 \, mg.l^{-1} \, H_2S$, generated by activity of sulphate-reducing bacteria, have been measured in locations such as drilling and utility legs on concrete platforms and on other installations. Such levels can lead to toxic concentration of H_2S in the atmosphere if precautionary measures are not taken. Successful control measures have been developed (Wilkinson, 1982) involving seawater exchange and sparging, and avoiding the use of biocides.

Corrosion

Once again sulphate-reducing bacteria are implicated although other micro-organisms may also be involved. This is an area where our understanding of the links between microbial activity and corrosion are developing rapidly at the present time (Gilbert *et al.*, 1984; Stranger-Johannessen, 1986). Much of the present symposium is centred on this area. There is considerable motivation for industry to understand the mechanisms involved and the factors contributing to corrosion, in order to establish relevant and adequate control measures. Biocide use can be justified on a careful selective basis, for example in shut-in injection wells and in some pipeline commissioning procedures.

Reservoir Souring

This subject is also addressed in detail elsewhere in the present symposium (Herbert, 1986). There is no doubt that microbial sulphate reduction can take place in closed-in injection wells and in pipelines filled with seawater, necessitating prophylactic biocide treatment. However, any activity deep in the reservoir will depend on a number of specific factors such as temperature and pH, extent of mixing etc. and there is a need to develop understanding of such physical and chemical parameters.

Marine Fouling

Although marine fouling is not regarded as a major structural problem for present structures, it may become of increasing significance in the design of future installations, particularly for small and marginal fields, and a thorough understanding of the contribution of growth to factors such as drag and structural loading (Wright and Bryce, 1983) will improve the accuracy of structural design tolerances. Recent evidence suggests that direct attachment to surfaces can involve a wide range of organisms and particulate matter, including many micro-organisms.

Microbial Contamination of Fuel

Problems of blockage of filters and corrosion of engine parts as a result of microbial growth in fuel have a long history in the shipping industry, not excluding shipping servicing the oil and gas industry. A detailed study of the latter has confirmed that contamination with water is the primary problem. In onshore storage tanks with a history of water contamination, much of the water has been shown to be seawater, leading to growth of sulphate-reducing bacteria and the associated possibility of corrosion. The most effective control measure is elimination of water; biocides should be used with extreme caution because of mixing with the fuel and associated health risks and because sloughing off of dead biomass can lead to exacerbation of the original problem.

Table 9   Estimated annual inputs of petroleum hydrocarbons to the North Sea* ('000 tonnes) (Institute of Offshore Engineering, 1985)

Natural seeps	0.3–0.8
Atmospheric	19
Rivers, land runoff (including inland municipal waste)	40–80
Coastal sewage discharges	3–14
Coastal refineries	6.0
Oil terminal operations (including reception facilities)	0.8
Other coastal industrial effluent	9
Accidental losses from tankers at sea	5–12
Operational discharges from tankers at sea	?
Losses from general shipping	?
Offshore production	23
Total	107–165

*North Sea defined as waters north of the Strait of Dover, the Skagerrak and the area to 61°N to a line at 4°W.

Table 10   Some inputs to the North Sea

	'000 tonnes.y^{-1}
Hydrocarbons — all countries	
Offshore oil and gas[a,b]	23
All sources[a]	107–165
Biological Oxygen Demand	
Total all countries[a]	800–1400
UK[a]	194–500
UK Offshore oil and gas[b]	80–93
Nitrogen	
Total all countries[a]	570–970
UK[a]	108–194
Offshore oil and gas[b]	1–2

[a]Institute of Offshore Engineering, 1985.
[b]Data from this paper.

OTHER INPUTS TO THE NORTH SEA

When placed in context with other hydrocarbon inputs to the North Sea (Table 9) inputs from the offshore oil and gas industry account for 20% or less of the total. Of the specific oil and gas inputs the major input is from drilling cuttings, a major part of which does not exert biological oxygen demand (BOD) in the short term.

Total biological demand from the industry's activities is of the order of 80,000 to 100,000 tonnes oxygen per year of which 20,000 to 25,000 tonnes arises from oil discharges and the remainder from other activities.

Some figures for comparison are shown in Table 10. For example, a BOD of some 210,000 tonnes per year is discharged from UK coastal sewage

Table 11   Inventory of data, North Sea oil activity (Read, 1985)

Operations as at end 1982	Denmark	W. Germany	Netherlands	Norway	UK	Total
No. of oil production platforms	3		2	14	31	50
Pipelines offshore (km) — oil	45		412	214	1,454	1,713
— gas	39	477	57	162	2,120	3,210
— other products		11		22	252	342
TOTAL	84	488	469	398	3,826	5,265
Wells offshore oil/gas total drilled by end 1982						
— production	48		104	236	1,099	1,487
— exploration & appraisal	50	51	319	360	965	1,745
TOTAL						3,232
Number of offshore loading points	1			1	7	9

Table 12   North Sea oil activity during 1985. (Read, 1985)

WELLS DRILLED — production	14	20	23	118	175
— exploration & appraisal	5 7	45	49	111	217
OIL PRODUCTION (tonnes)	2,000,000	150,000	24,500,000	103,300,000	129,950,000
GAS PRODUCTION (MCM)		11,000	24,000	38,000	73,000
OIL TRANSPORTED — by pipeline		150,000	13,200,000	92,400,000	105,750,000
— offshore loading	2,000,000		11,300,000	10,900,000	24,200,000
ACCIDENTAL SPILLS					
offshore installations — number	—	—	12	42	54
— total tonnes	—	—	165	162	327
crude loading terminals (total tonnes)	—	—	<150	<150	<150
OPERATIONS DISCHARGES — tonnes oil in					
Produced water	11	—	253	927	1,191
Cuttings from oil-based mud usage — total oil	130	NA	c. 40	c. 500	c. 540
Drainage water	0.2	NA	2–5	10–20	12–25
Terminal Effluents	—	—	—	400	400
DEBRIS DAMAGE CLAIMS — number	10	—	343	67	420
— total compensation	D.Kr.600,000		N.Kr.6,000,000		
	£45,000		£545,000	£107,000	£697,000

outfalls as part of a total BOD of 800,000 to 1,400,000 tonnes discharged to the North Sea which includes the oxygen demand exerted by 86,000–142,000 tonnes of oil (excluding oil from the offshore industry).

Thus, although inputs from the offshore operations are significant they are small relative to other inputs. Moreover, inputs are from a large number of dispersed sources as indicated by reference to Figure 1, which shows all drilling locations for Shell Expro since the commencement of exploration in the North Sea. Although confined to Shell Expro licence areas the pattern will be similar for the overall UK sector activities. From this it can be seen that discharges are made from a large number of dispersed locations. This is particularly relevant to drilling mud discharges for which investigations to date have concentrated on multi-well production platforms. It can be predicted that effects around single-well sites will be much less than for multi-well sites and recovery much more rapid, particularly where debris trawling is carried out as discussed above.

Overall North Sea offshore oil and gas operations are summarised in Tables 11 and 12 (1982 data) (Read, 1985), clearly indicating that the UK is the major national source of activity.

SUMMARY AND CONCLUSIONS

There is much detailed knowledge of the quantity of emissions to the North Sea from the offshore oil and gas industry. This is particularly true of the hydrocarbons discharged from drilling and production operations.

Although the quantities of hydrocarbons and other components giving rise to oxygen demand in the sea are significant, these are relatively small and widely dispersed compared to some land-based sources.

There is no evidence of unacceptable damage to the environment arising from offshore operations. This has recently been confirmed by detailed scientific consideration of the environmental effects of North Sea oil and gas developments (Royal Society, 1986).

Nonetheless there is no reason for complacency. Continued vigilance will be required, most appropriately through a combination of sound analyses and quantification of discharges with knowledge of their dispersion in the environment and any biological effects. The economic incentive to minimize oil losses is an additional and ongoing motivation.

The capacity of micro-organisms to metabolize particular compounds and assimilate them into the food chain is an important aspect of the environmental interactions. However, in this area, much is predictable at least in broad terms, based on previous work on microbial metabolism. Thus, it can be assumed that much of the organic matter discharged is readily metabolized.

Micro-organisms can also have adverse effects on oil and gas operations including involvement in corrosive processes and deterioration of stored fuel.

In the offshore environment this is particularly true of the activities of sulphate-reducing bacteria. There is an ongoing need to improve multi-disciplinary understanding of some of the contributions of micro-organisms which may have direct economic implications.

REFERENCES

Bedborough, D. R. and Blackman, R. A. (1986). 'A survey of inputs to the North Sea resulting from oil and gas development', *Proc. R. Soc. Lond.*, (in press).

Davies, J. H., Addy, J. M., Blackman, R. A., Blanchard, J. R., Ferbrache, J. E., Moore, D. C., Somerville, H. J., Whitehead, A. and Wilkinson, T. (1984). 'Environmental effects of the use of oil-based drilling muds in the North Sea', *Marine Pollution Bulletin*, **15**, 363–370.

Department of Energy (1979). Continental Shelf Operations Notice 11. 'Consents to the drilling and abandonment of wells'.

Department of Energy (1981). Continental Shelf Operations Notice 8. 'Loss or dumping of synthetic materials or other refuse at Sea'.

Department of Energy (1985). 'Development of the oil and gas resources of the United Kingdom 1985', London, HMSO.

Gilbert, P. D., Steele, A. D., Morgan, T. D. B. and Herbert, B. N. (1984). 'Microbial problems and corrosion in oil and oil product storage', *Institute of Petroleum Symposium 1983*, pp. 71–80. London, Institute of Petroleum.

Goodman, K. S. and Troake, R. P. (1984). 'Environmental effects of production water discharges', *Petroleum Review*, August, 1984.

Herbert, B. N. (1986). '*Reservoir souring*', this symposium.

Higgins, I. J. and Gilbert, P. D. (1978). 'The biogradation of hydrocarbons', in *The Oil Industry and Microbial Ecosystems* (eds. Chater, K. W. A. and Somerville, H. J., pp. 80–117. London, Heyden and Sons.

Institute of Offshore Engineering (1985). 'Input of contaminants to the North Sea from the United Kingdom', (compiled by W. C. Grogan). Final report prepared by the Institute of Offshore Engineering, Heriot-Watt University, for Department of Environment. Edinburgh, Institute of Offshore Engineering.

Paris Commission (1984). Sixth Annual Report. London, Paris Commission.

Prandle, D. (1984). 'A modelling study of the mixing of ^{137}Cs in the seas of the European Continental Shelf', *Phil Trans. R. Soc. Lond.*, **A310**, 407–436.

Read, A. D. (1985). 'Oil Industry collaboration in the monitoring and protection of the North Sea', Greenwich Forum XI. London, Oil Industry Exploration and Production Forum.

Royal Society (1986). 'Environmental Effects of North Sea Oil and Gas Developments', (in press).

Somerville, H. J. (1984). 'North Sea exploration and production: interactions with the environment', Proceedings, Institute of Petroleum Annual Conference 1984, *Offshore U.K* pp. 100–122. London, Institute of Petroleum.

Somerville, H. J. (1985). 'Physiological aspects of biotreatment of petrochemical wastes', *Conservation and Recycling*, **8**, 73–83.

Stranger-Johannessen M. (1986). 'Microbial deterioration of corrosion protective coatings', this symposium.

Vase, E. J. (1983). 'The role of chemicals in oil and gas production', in *Chemicals in the Oil Industry*, (ed. Ogden, P. H.), pp. 61–72. London, Royal Society of Chemistry.

Watkinson, R. J. (1978). 'Developments in biogradation of hydrocarbons', London, *Applied Science*.

Wilkinson, T. G. (1982). 'An environmental programme for offshore oil operations', *Chemistry and Industry*, Feb. 10, 1982, 115.

Wright, R. A. D. and Bryce, A. A. (1983). 'Assessment of marine growth on offshore installations', *Proceedings Offshore Europe 1983*, 379–390. Soc. Petr. Eng. Paper SPE 1190411.

ENVIRONMENTAL SURVEYS TO STUDY
THE RECOVERY OF MARINE SEDIMENTS
CONTAMINATED BY OILED DRILL CUTTINGS

L. C. Massie and J. M. Davies

DAFS Marine Laboratory, Aberdeen, Scotland

ABSTRACT

At the present time the largest contribution to the total (44,000 tonnes per annum) hydrocarbon input to the North Sea is 20,000 tpa oil attached to drill cuttings. The immediate effect of these oiled cuttings on the sediments and benthic communities close to platforms has proved to be severe, and information on their residence time has been required. To this end the recovery of the sediments at two platforms has been studied in detail by monitoring the relative changes in hydrocarbon chemical composition with time and distance from platforms. Attention has been paid to the depth of penetration of hydrocarbons into the sediment and to the potential of the indigenous microbial populations to mineralize these hydrocarbons.

INTRODUCTION

UK continental shelf waters support some of the most productive and successful fisheries in the world. It is important therefore to safeguard fishing interests and to preserve the North Sea and other UK waters for fisheries of the future. It is equally important to allow other legitimate users of the seas (e.g. shipping, sewage dumping, mineral extraction) to operate with the minimum of hindrance and to reassure our neighbouring countries that we manage our various operations in such a way that we pose no threat to their interests.

The arrival of the oil industry to UK waters brought worries to the fishing industry. Loss of access to fishing grounds, damage to fishing gear from oil-related debris, and damage to fish and shellfish stocks due to pollution all gave cause for concern. To date exploration and production of oil and gas has been concentrated mainly in the North Sea. Hydrocarbon input to this area is presently estimated to total 44,000 tonnes per annum (tpa) (based upon the 8th Royal Commission on Environmental Pollution, 1981),

125

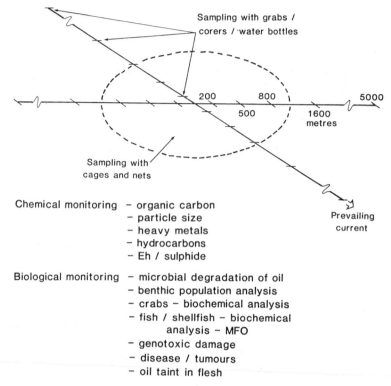

FIG 1   Complete monitoring programme used around oil production platforms.

half from sources such as shipping, refineries, tankers, sewage, land run-off and the atmosphere and the remainder from the oil industry — 2,000 tpa oil in production water and 20,000 tpa oil attached to drill cuttings from platforms using oil-based drilling muds (Blackman and Bedborough, 1986). The latter source has increased over the years from 1977 to its present level though since 1981 the type of oil used has changed from diesel to less acutely toxic paraffinic/naphthenic oils.

In response to the increasing use of oil-based mud the government and industry mounted monitoring programmes around platforms, to determine the environmental impact of these muds. It was soon established (Davies *et al.*, 1984) that there was a very steep hydrocarbon gradient around those platforms using oil-based drilling muds, with higher than background levels extending 2–3 km from the platform, and contamination sometimes extending as far as 5–8 km in the direction of the prevailing current. Serious biological effects within the sediment animal community were observed within 0.5–1 km of the platforms, returning to normal outwith this area. It is important to know how quickly these contaminated areas around platforms will recover when drilling is finished. To this end a study was conducted around the Beryl 'A' platform where diesel-based mud had been used extensively and also around the Beatrice 'A' platform where low toxicity oil-based muds were used.

The factors which will reduce the hydrocarbon concentrations in surface sediments are lateral movement, vertical mixing and biodegradation. The data collected has been examined with these in mind.

METHODS

The complete monitoring programme is summarized in Figure 1. For this study results will be presented for redox potential (Eh), organic carbon and hydrocarbon concentrations, and microbial mineralization rates only.

Sediment samples were taken with a grease-free Smith-McIntyre grab at the various sampling stations (Figure 1), the upper 1–2 cm layer of sediment only being used for chemical analysis and for measurement of

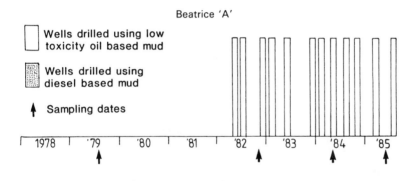

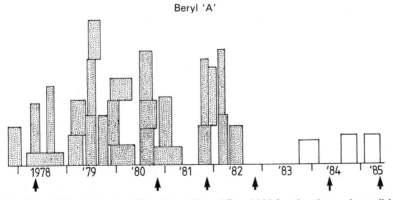

FIG 2   The drilling history of Beatrice 'A' and Beryl 'A' for the time when oil-based drilling muds were in use. Each well represents the addition of 90 tonnes of oil to the sediments.

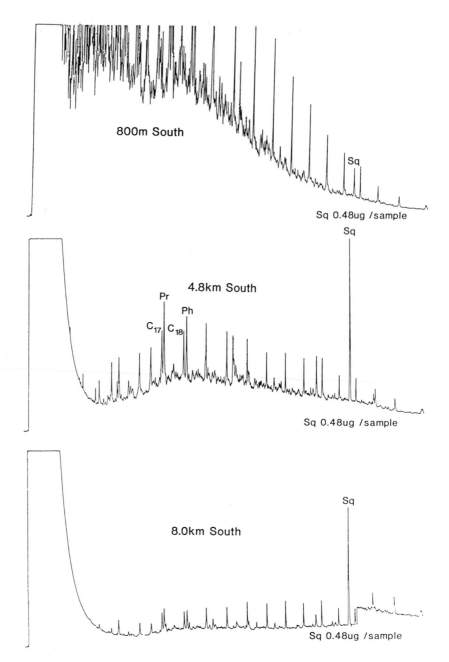

FIG 3   Gas chromatographs of aliphatic hydrocarbons of base oils from freshly drilled cuttings.

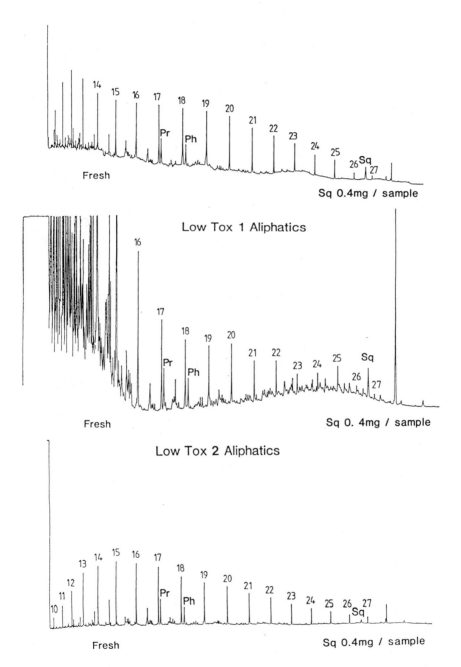

FIG 4   Typical gas chromatographs of aliphatic hydrocarbons found in surface sediments at Beatrice 'A' 1984.

heterotrophic activity using 1-^{14}C naphthalene, 1-^{14}C hexadecane and ^{14}C (U) amino acids. Core samples were taken by a Craib corer or by a multi-corer, some cores being used for Eh measurements by the method of Pearson and Stanley (1979) and others divided into 0–2 cm, 2–4 cm and 4–8 cm portions for chemical analysis.

Hydrocarbons were extracted from the sediments using dichloromethane. After all the extracts had been screened by UV fluorescence measurements against standard dilutions of diesel or low toxicity base oil, certain samples were selected for more detailed analysis by capillary GLC and/or GC-MS (Massie et al., 1985a).

1-^{14}C naphthalene was used to measure the potential of indigenous microbial populations to degrade oil present in sediments since field and laboratory experiments had shown that, of the limited ^{14}C 'oil' substrates available, it was the most readily mineralized (Massie et al., 1985b). Around Beatrice 'A' where low toxicity oils were used, 1-^{14}C hexadecane mineralization rates were also measured although tests had shown it to have a low potential. The rate of mineralization of ^{14}C (U) mixed 1-amino acids was used as a measure of total heterotrophic activity. The method is detailed in Saltzmann (1982) and incubation times used were 2 hours for ^{14}C (U) amino acids and 24 hours for 1-^{14}C hexadecane. For 1-^{14}C naphthalene the incubation time was also 24 hours except with sediments from stations close to the Beryl platform where mineralization rates were expected to be high. At these the time was reduced to 12 hours.

Organic carbon was estimated, in a Perkin Elmer CHN analyser, on some of the samples which had been freeze dried and freed of carbonate by hydrochloric acid treatment.

RESULTS

The drilling histories for the time of use of oil-based muds at the two platforms under consideration are shown in Figure 2. It is assumed that each well contributed about 90 tonnes oil (attached to 510 tonnes cuttings) to the sediments (Blackman and Bedborough, 1986).

For later comparison Figure 3 shows capillary gas chromatographs typical of the aliphatic hydrocarbons found in fresh diesel cuttings and also fresh cuttings from the use of two different low toxicity base oils. The important factors to note are the size and shape of any UCMs present and the nC_{17}/pristane and nC_{18}/phytane ratios (Blumer et al., 1973) which are close to 2.

Lateral Movement of Oil Around Platforms

Figure 4 shows some typical chromatographs of aliphatic hydrocarbons found close to Beatrice 'A' (Figure 4(a)) and at increasing distances (Figures

4(b) and 4(c)) from the platform. There was evidence of low-level hydrocarbon contamination as far as 8 km to the south and west and 5 km to the east of the platform, as shown by slight UCMs and by the nC_{17}/pristane and nC_{18}/phytane ratios being close to 1. At intermediate stations UCMs were more noticeable (Figure 4(b)) and within 2 km north, south and west and 1 km east they were pronounced and increased with decreasing distance from the platform. The nC_{17}/pristane and nC_{18}/phytane ratios were close to 1 at stations 0.4 km north and east and 0.4-3.2 km south. At all the other intermediate stations the ratios were in the range 0.4-0.7.

Aromatic hydrocarbon fractions of extracts from Beryl 'A' sediments were analysed by GC-MS for selected 2-6 ring compounds (for the detailed list see Massie et al., 1985a). Their relative composition and total concentrations are shown in Figure 5. In 1980, after 2½ years of continuous drilling with diesel-based muds, the sediments sampled 0.8 km south contained oiled cuttings which were relatively fresh since their aromatic profile was very similar to that found in diesel. At the 1.6 km station, however, most of the naphthalenes had been preferentially removed leaving mainly the 3-ring phenanthrenes and anthracenes with their C_1-C_4 alkyl derivatives and also the dibenzthiophenes. The relative composition of the aromatic hydrocarbons within 5 km south of the platform did not show a profile similar to that found at 'clean' stations although the total concentrations had fallen to 'clean' or 'background' levels by 3.2 km south.

In 1985, approximately 3 years after drilling with diesel-based mud had ceased and only 3 further wells had been drilled with low toxicity oil-based muds (Figure 2), hydrocarbons associated with oiled cuttings were found 8 km south of the platform. The relative composition of the aromatic hydrocarbons showed that naphthalenes comprised 50% of the total at the 0.2 and 0.5 km stations falling to 25% at 5-8 km distance (Figure 5). The evidence suggests that there is a time dependent biodegradation with distance from the platform.

Vertical Mixing

In 1984, 2 years after drilling with diesel-based mud had ceased and immediately after completion of the first well using low toxicity oil-based mud (Figure 2), analysis of cores from the Beryl 'A' platform by capillary gas chromatography showed that large UCMs were present in the 0-2 cm depth section at the close stations but the contamination decreased rapidly with increasing depth in the sediment (Figure 6). Preferential removal of lower molecular weight alkanes was found in the 0-2 cm sections (as typified by Figure 6(a)) and in the 2-4 cm sections (Figure 6(b)) but in the 4-8 cm sections there were no UCMs on the baselines and the most

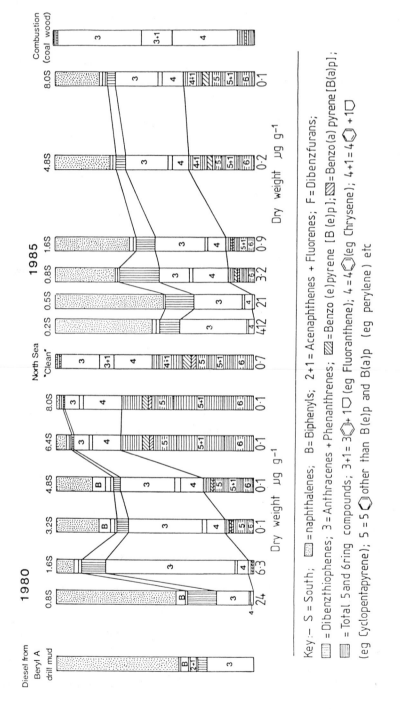

FIG 5   The relative composition and total concentration of selected aromatic hydrocarbons in surface sediments at Beryl 'A' 1980, 1985.

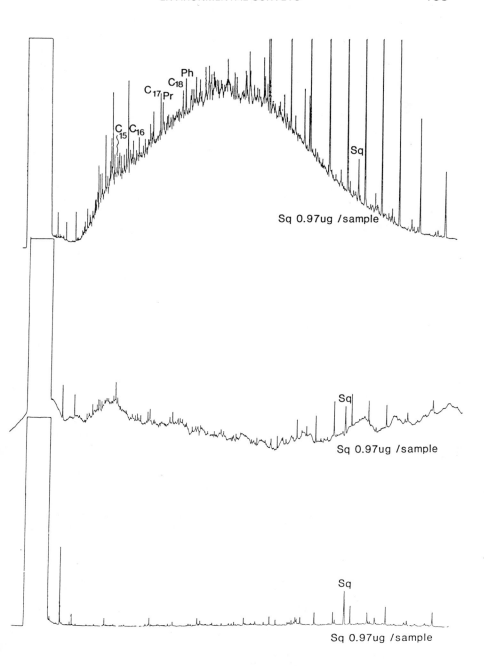

FIG 6   Gas chromatographs of aliphatic hydrocarbons in a core sample from 0.2 km south of Beryl 'A' 1984.

prominent alkanes present were the nC_{27}, nC_{29} and nC_{31} homologues which were present in much higher levels than their corresponding nC_{28} and nC_{30} homologues (Figure 6(c)). This alkane profile is typical of terrestrial plants and has been seen in sediments in many stations in the North Sea (Hardy *et al.*, 1977). This suggested that at stations outwith 0.2 km from the platforms there was relatively little penetration of the hydrocarbons into the sediments.

GC-MS analysis of some of the 1984 cores also showed that, at the close station (0.2 km south), total aromatic hydrocarbon concentrations decreased with increasing depth in the sediment (Figure 7). Their relative composition showed that partially degraded diesel-type aromatic compounds were not present below 4 cm since the 4–8 cm section showed a profile typical of 'background' sediment. At the more distant station (1.6 km south-east) the total aromatic concentrations were uniform with depth into the core sample and in both concentration and composition were similar to background.

Biodegradation

Total organic carbon and redox potential

The total organic carbon concentrations in the sediment samples taken in the Beryl field in 1984 and 1985 are shown in Table 1 together with the

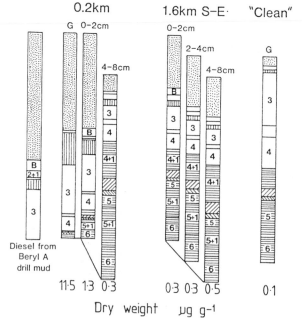

FIG 7   The relative composition and total concentration of selected aromatic hydrocarbons in cores at Beryl 'A' 1984.

Table 1　Redox potential in cores (2.5 cm and 4.0 cm) and carbon content of surface sediments at Beryl 'A' 1984, 1985

	1.6 N	0.8 N	0.5 N	0.2 N	0.2 S	0.5 S	0.8 S	1.6 S	3.2 S	4.8 S	0.2 SE	0.8 SE	'Clean'
1984													
%C	0.38			0.73	0.97	1.01	0.62	0.65	0.42	0.48			0.30
Eh (mV)													
2.5 cm			+281		+68		+415			+424		+428	+189
4 cm			+249		+45		+392			+380		+416	+118
RPD (cm)			>10		9.50		>10			>10		>7	>10
1985													
%C	0.43			0.76	1.24	0.67	0.47						
Eh (mV)													
2.5 cm		+151	+120	+195		+4	+106		+286	+157	−85	+296	
4 cm		+103	+58	+83		−80	+81		+204	+121	−150	+216	
RPD (cm)		>10	>8	5.17		2.56	>10		>10	>10	0.90	>10	

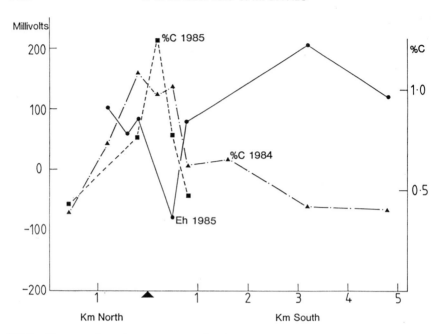

FIG 8   The organic carbon content and redox potential (Eh) of sediments at Beryl 'A'.

Eh results expressed in millevolts (corrected) at 2.5 cm and 4.0 cm depth in the sediments. The redox potential discontinuity (RPD) depths (i.e. the depth at which the voltages become zero) are also included. Some of the results are shown graphically in Figure 8.

The redox potential of the sediments decreased with increasing carbon concentrations in both surveys. The lowest values found in 1985 were at stations 0.2 km south-east where the RPD was close to the surface (0.9 cm) and 0.5 km south (RPD = 2.56 cm). Addy *et al.*, (1984) also found reduced conditions in sediments at 2.5 cm depth within 0.2 km of the Beatrice 'A' platform after 5 wells had been drilled with low toxicity oil-based muds.

The redox values measured at the stations where heterotrophic measurements were made showed that the surface sediment used for these assays had positive Eh values.

Mineralization of C¹⁴ Labelled Hydrocarbon Substrates

The mineralization rates measured for 1-^{14}C naphthalene (potential mineralization rates) at Beryl and Beatrice are shown in Figure 9. At Beryl 'A' the highest rates were always found closest to the platform, the maxima being 3.1 mg/m²/cm/day (1980) and 3.0 mg/m²/cm/day (1982). At stations beyond 1 km mineralization rates were usually higher north and

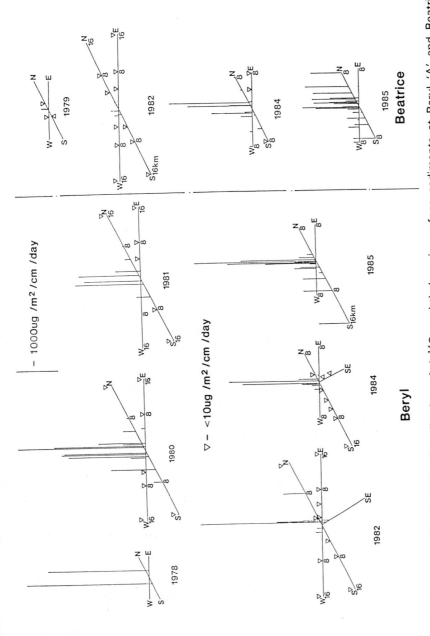

FIG 9   The rates of microbial mineralization of 1-^{14}C naphthalene in surface sediments at Beryl 'A' and Beatrice 'A' 1978-1985.

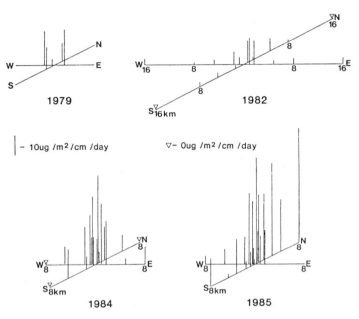

FIG 10   The rates of microbial mineralization of $1\text{-}^{14}C$ hexadecane in surface sediments at Beatrice 'A' 1979–1985.

south of the platform than east and west and the year to year variation in rates at these stations appeared to reflect the immediate oil-based drilling history of the platform. For example the lowest rates (often zero) for the Beryl stations were found in 1982, several months after drilling with diesel had ceased and little increase was seen in 1984 immediately following the completion of one well using low toxicity oil-based mud. In 1985, however, after an additional 2 wells had been drilled with low toxicity oil-based mud the increase in mineralization rate was marked. Close to the platform the highest rate found ($2.9\,\text{mg/m}^2/\text{cm/day}$) was similar to the values found in 1980 and 1982.

At Beatrice 'A' low potential mineralization rates were found for naphthalene in 1979 and also in 1982 after 13 wells had been drilled with water-based muds and 2 wells with low toxicity oil-based mud had been completed. However, 2 years and 6 further oil-based mud wells later (Figure 2), there was marked increase in 1984 with a maximum rate of $1.8\,\text{mg/m}^2/\text{cm/day}$ being found close to the platform and degradation taking place out to 5 km south, west and north. In 1985 after a total of 11 wells had been completed with the use of low toxicity oil-based mud mineralization rates had increased yet again at the further stations but not close to the platform.

The potential mineralization rates for hexadecane at Beatrice 'A' (Figure 10) showed a similar response timewise to the addition of oiled cuttings

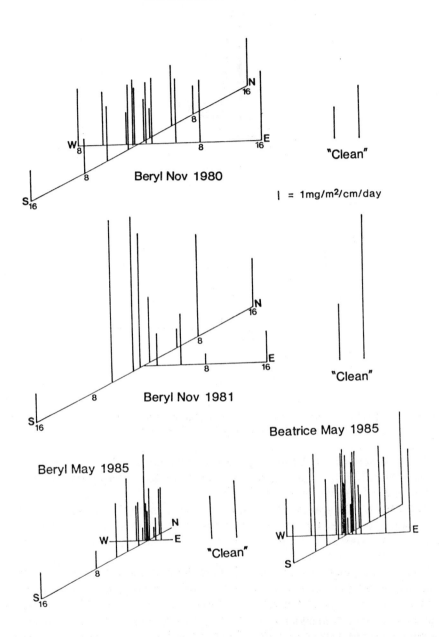

FIG 11 The rates of microbial mineralization of 1-amino acids in surface sediments at Beryl 'A', Beatrice 'A' and at 'clean' (remote from oil activity) stations calculated on the assumption that 10,000 μg/m^2/cm amino acids are present in surface sediments (from Henricks and Farrington, 1979).

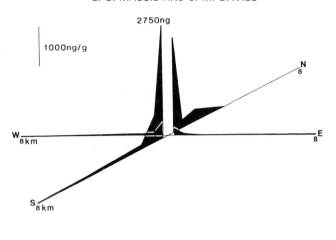

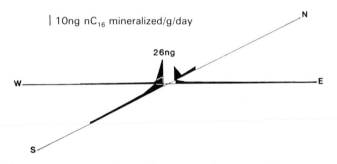

FIG 12   The concentration and actual mineralization rates of hexadecane in surface sediments at Beatrice 'A' 1984.

to the sediments as that found for naphthalene except that the rates were 1–2 orders of magnitude lower, the maximum being $45\,\mu g/m^2/cm/day$ in 1984 and $57\,\mu g/m^2/cm/day$ in 1985.

The rate of mineralization of amino acids (Figure 11) appeared to be remarkably uniform around the platforms indicating an adaption of, rather than an increase in, the indigenous microbial populations to the addition of oil to the sediments.

The actual mineralization rates of hexadecane and naphthalenes were calculated from the concentrations of these substrates measured in sediments and the percentage of added $1\text{-}^{14}C$ hexadecane and $1\text{-}^{14}C$ naphthalene mineralized per day in sediments from the same grab sample. The results for hexadecane at Beatrice in 1984 (Table 2 and Figure 12) showed that the actual mineralization rates reflected the concentrations of hexadecane found in the sediments. This was true also for the mineralization rates for the naphthalenes at Beryl 'A' in 1985 which are shown in Table 2 and Figure 13.

Table 2   Actual mineralization rates of naphthalenes and hexadecane in surface sediments at Beryl 'A' (1985) and Beatrice 'A' (1984)

	Beryl 1985			Beatrice 1984		
	Naphths ng/g dry wt	$\%^{14}CO_2$ per day	Naphthalenes mineralized ng/g/day	nC_{16} ng/g dry wt	$\%^{14}CO_2$ per day	nC_{16} mineralized ng/g/day
1.6 km N				291	1.16	3.38
0.8	194	12.0	23.4	163	0.36	0.59
0.5	1811	2.3	41.7			
0.4				2378	0.61	14.51
0.2	1552	46.3	718.6			
0.2 km S	203617	39.3	80021.4			
0.4				2755	0.95	26.17
0.5	11944	58.9	7035.0			
0.8	986	39.2	386.5	653	0.92	6.01
1.6	347	9.9	34.4	112	0.39	0.43
4.8	51	14.7	7.5	15	0.94	0.14
8.0	29	18.5	5.4	5	0	0
8.0 km W				8	0	0
4.8				6	0.62	0.04
1.6				39	1.30	0.51
0.8				40	1.69	0.67
0.4				292	1.86	5.41
0.2	89	15.3	13.6			
0.2 km E	1352	14.9	201.4			
0.4				151	2.62	3.96
0.8				28	1.64	0.46
1.6				7	1.53	0.10
4.8				3	0.14	+
8.0				31	0.63	0.20
0.2 km SE	18342	4.4	807.0			
0.8	114	7.7	8.8			

DISCUSSION

Comparison of gas chromatographs of the aliphatic fractions of the oils from fresh cuttings (Figures 3(b) and (c)) compared with those from sediments around the Beatrice 'A' platform (Figure 4) showed that the smaller n-alkanes (nC_{10}–nC_{20}) were being preferentially removed. The isoprenoid compounds pristane and phytane were degraded more slowly and the alicyclic (naphthenic) compounds were accumulating in the UCMs. Tibbetts and Large (in press) found similar results at the same platform and also found a ring of intermediate stations where the nC_{17}/pristane and

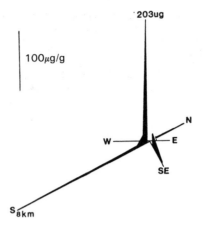

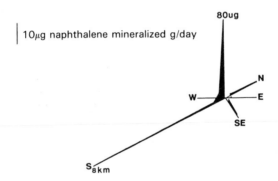

FIG 13   The concentration and actual mineralization rates of naphthalenes in surface sediments at Beryl 'A' 1985.

nC_{18}/phytane ratios were much lower than those found closer to and further from the platform. They suggested that when concentrations of n-alkanes occurred within certain limits in the sediments then their optimum degradation rates could be achieved. However, the potential and actual mineralization rates found for hexadecane (nC_{16}) did not confirm this but showed rather that the rate increased with increasing concentrations of the alkane in the sediment (Figures 10 and 12). Very possibly, as was suggested by Tibbits and Large, there is a level below which heterotrophs will not readily degrade hexadecane (and similar n-alkanes) if other, more suitable, substrates are available. This would explain the finding that nC_{17}/pristane and nC_{18}/phytane ratios were close to 1 at the further stations, rather greater than those found at the intermediate stations (0.4–0.7). At stations close to the platform the true degradation rates were probably masked by the continuous input of fresh cuttings (with a ratio of 2), increasing the ratios nearer to 1.

The aromatic profiles found at Beryl in 1985 showed the changes that had occurred since 1980 at comparable stations. In 1980 there was evidence of diesel, which was fresh at the close stations (Figure 5), degraded at intermediate stations and absent beyond 5 km south. In 1985, 3 years after intensive drilling with diesel had ceased and 3 wells had been drilled with low toxicity base oil, there was evidence of the oil contamination spreading as far as 8 km south. At stations closer than 0.8 km south the input of aromatic compounds from the low toxicity oil was evident in the increased proportion of naphthalenes and 3-ring compounds present and the decreasing proportion of dibenzthiophenes. The combined evidence indicated the presence of diesel residues and a fresh input of naphthalenes from the 3 wells drilled with low toxicity muds out to 8 km south from the platform.

A similar picture emerged from a study of the microbial activity. After drilling with diesel had ceased in July 1982 the potential mineralization of naphthalenes had, by November 1982, returned to background rates at stations outwith 0.5–1 km distance and remained so in 1984 but in 1985 when the chemical measurements showed a fresh input of naphthalenes to be present the rates had increased substantially (Figure 9). Although the levels of naphthalenes were very low in the surface sediments 8 km south of the platform in 1985 they were still sufficient to induce higher rates of naphthalene mineralization.

A rough estimate of the time required by the heterotrophs to degrade naphthalenes around platforms can be obtained for sediments where aerobic conditions prevail and by using the fact that decreasing concentrations are mineralized at decreasing rates (Table 2). Allowing for this, the time taken for naphthalenes to be mineralized to background levels was calculated to be 1–2 months. Good agreement with the estimate was found in 1982 when very low or zero potential mineralization rates were found for naphthalene at stations outwith 0.5 km of the Beryl 'A' platform, three months after drilling with diesel had ceased.

In a similar manner the time required for the mineralization of hexadecane at Beatrice was calculated and found to be 3–6 months for moderate levels (up to 600 μg/g dry weight sediment) and up to one year for the high levels ($>$ 2000 μg/g dry weight sediment) present 0.4 km north and south of the platform.

The removal times calculated for naphthalenes and hexadecane cannot be applied to all hydrocarbon compounds present in the sediments. Experiments with larger polyaromatic compounds showed that their mineralization potentials were 2–3 orders of magnitude lower than that found for naphthalene (Massie et al., 1985b). Also the chemical evidence showed the accumulation of these refractory aromatic (and also aliphatic) compounds in the sediments. They will apparently remain for many years after drilling has ceased.

CONCLUSIONS

Lateral movement of oiled cuttings, presumably mediated by currents, was detected as far as 8 km distance from both platforms. Vertical mixing of the oil in the sediments outwith 0.2 km of the platforms was not extensive. In the samples examined, evidence of oil was not found below 4 cm depth.

The rate of mineralization of naphthalenes and hexadecane reflected the concentration of these compounds in the sediments. The removal times calculated were 1–2 months for naphthalenes and up to one year for hexadecane in fully aerobic sediments.

The chemical evidence confirmed that the smaller n-alkanes and aromatic hydrocarbons were being preferentially removed from the oiled sediments leaving the refractory compounds to accumulate. They are likely to remain in the sediments for many years after drilling with oil-based muds has ceased.

REFERENCES

Addy, J. M., Hartley, J. P. and Tibbetts, P. J. C. 'Ecological effects of low toxicity oil-based mud drilling in the Beatrice oilfield', (in press).

Bedborough, D. R. and Blackman, R. A. A. 'A survey of inputs to the North Sea resulting from gas and oil developments', *Phil. Trans. Royal Soc. London, Series B* (in press).

Blumer, M., Erhardt, M. and Jones, J. H. (1973). 'The environmental fate of stranded crude oil', *Deep Sea Research*, **20**, 239–259.

Connan, J. (1986) 'Biodegradation of crude oil in the reservoir', (this volume pp. 49–56).

Davies, J. M., Addy, J. M., Blackman, R. A., Blanchard, J. R., Ferbrache, J. E., Moore, D. C., Somerville, H. J., Whitehead, A. and Wilkinson, T. (1984). 'Environmental effects of the use of oil-based drilling muds in the North Sea', *Mar. Pollut. Bull.*, **15**, 363–370.

Hardy, R., Mackie, P. M., Whittle, K. J., McIntyre, A. D. and Blackman, R. A. A. (1977). 'Occurrence of hydrocarbons in the surface film, sub-surface water and sediments in the waters around the United Kingdom', *Rapp. P. -v. Reun. Cons. int. Explor. Mer*, **171**, 61–65.

Henricks, S. M. and Farrington, J. W. (1979). 'Amino acids in interstitial waters of marine sediments', *Nature*, **279**, 319–321.

Massie, L. C., Ward, A. P., Davies, J. M. and Mackie, P. R. (1985a). 'The effects of oil exploration and production in the northern North Sea: Part 1 — The levels of hydrocarbons in water and sediments in selected areas, 1978–1981', *Mar. Environ. Res.*, **15**, 165–213.

Massie, L. C., Ward, A. P. and Davies, J. M. (1985b). 'The effects of oil exploration and production in the northern North Sea: Part 2 — Microbial biodegradation of hydrocarbons in water and sediments, 1978–1981', *Mar. Environ. Res.*, **15**, 235–262.

Pearson, T. H. and Stanley, S. O. (1979). 'Comparative measurements of redox potential of marine sediments as a rapid means of assessing the effect of organic pollution', *Mar. Biol.* **15**, 371–379.

Royal Commission on Environmental Pollution. Eighth Report, 'Oil Pollution of the Sea', (1981). HMSO, London.

Saltzmann, H. A. (1982). 'Biodegradation of aromatic hydrocarbon in marine sediments of three North Sea oil fields', *Mar. Biol.*, **72**, 17–26.

Tibbetts, P. J. C. and Large, R. 'The degradation of low toxicity oil-based drilling mud in benthic sediments around the Beatrice oilfield', in *Marine Environment of the Moray Firth, The Royal Society of Edinburgh Proceedings, Series B* (in press).

THE USE OF ENVIRONMENTAL AUDIT IN OFFSHORE OPERATIONS

C. S. Johnston, J. C. Side and S. R. H. Davies

Institute of Offshore Engineering, Heriot-Watt University

BACKGROUND

Integral to the development and management of an offshore oil facility is a policy which ensures adequate safeguards for the protection of the environment. These safeguards must be derived from an understanding of the development and the possible risks presented to the environment; and of course appropriate understanding of the potential receiving environment. Any environmental protection policy must obviously include practical and cost-effective strategies both for monitoring operations themselves, and their possible impact on the receiving environment. Evidence of effects of a development on the environment must emerge at the earliest opportunity if it is to play a realistic part in the management of the operation.

This paper focuses on truly offshore oil developments in UK waters, where until recently there have been no statutory requirements for operators to assess impact of operations on the marine environment, a policy based largely on the assumption that the offshore North Sea has near infinite diluting power. The only control relevant to the protection of the marine environment was the control of oily water discharges. Within the exemption conditions from the prohibition of oil discharges from offshore production facilities as determined under the Prevention of Oil Pollution Act 1971, discharge limits were set based on best practical means for oil:water separation technologies (CUEP, 1976). These limits therefore were based on equipment performance rather than environmental criteria, although some later attempts have been made to extrapolate to possible environmental requirements (Read and Blackman, 1980).

In the inshore situation, for oil-loading terminals e.g. Flotta (Orkney) and Sullom Voe (Shetland) similar oily water discharge limits were set, under exemption conditions from the Prevention of Oil Pollution Act 1971; but these included environmental monitoring requirements e.g. 'that the operator shall ascertain the state of the flora and fauna within 2 kilometres of the discharge'. It might have been more prudent to have set some similar environmental monitoring requirement offshore and to have placed a

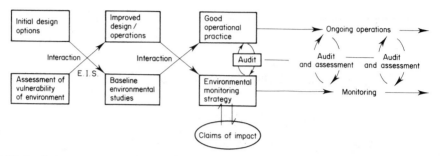

FIG 1   Relationship between EIA, ongoing environmental audit and monitoring.

responsibility upon an operator to demonstrate an environmental protection policy, appropriate to the location and development.

With no statutory demands, offshore environmental programmes in the UK North Sea developed within corporate company strategies, ranging from nil to detailed studies. Environmental criteria have been included in new statutory controls over the use of oil-based drilling muds, and these have had to be incorporated into company programmes, where these existed.

ENVIRONMENTAL IMPACT ASSESSMENT (EIA) AND AUDIT

The first steps in environmental protection come at the project proposal and design stage, when:

(i)   potential sources of risk to the environment, and
(ii) sensitive and vulnerable components within the receiving environment,

are recognized and quantified, and an appropriate strategy developed. This process of Environmental Impact Assessment (EIA) is a formal requirement upon developing industry in many countries, and will now be required for certain developments as a result of the EEC Council Directive (OJ L 175, 5/7/85), but is not mandatory (or likely to be) for offshore oil developments in the UK. It is, however, a useful procedure and is increasingly being employed by industry, but requires a solid technical base, and must be strongly supported by senior management, if it is to be successful.

Figure 1 identifies some of the key steps in a typical EIA, with initial project and environmental assessments defining in broad terms the interaction potential, often presented as an interaction matrix. The assessment focuses on key areas of potential interaction, establishing where possible quantitative bases for possible project risk sources and sensitive environmental components. From this comes the quantitative impact evaluation and possible mitigative measures.

Obviously, as the project moves forward to the final design and construction phases, technical changes will occur, as will planned operational practice, it is therefore essential that a full revision of the EIA is undertaken as the development goes into operation. This is really the first step, in what must be an ongoing audit process. As already stated, it is essential that the EIA undertaken for an offshore development is a thorough technical process with detailed engineering design data forming the basis — not the 'PR centred' glossies which appear as EIA documents in some industries. Consequently, the first AUDIT serves as a check on original assumptions/predictions made in the EIA and associated claims for plant performance from design engineers and manufacturers.

As the AUDIT proceeds, reliable records of process upsets prove especially useful, in particular for assessment of the effects of overloading on downstream processes. All process and operational changes (and their bases) must be carefully documented. Chemical usage patterns are notoriously difficult to predict, and downstream consequences can be complex, expensive and often unnecessary.

The safety or technical audit is a common practice in the oil industry, and a close relationship must be maintained between environmental and safety audits. This is of obvious importance in such matters as chemical usage (including drilling fluids).

Maintaining an ongoing audit of potential environmental risk ensures adequate and relevant:

(i)   process design and operational control;
(ii) performance/emission monitoring;
(iii) environmental (effects) monitoring.

Emission monitoring within several company environmental control strategies focuses on:

(i)   drilling fluid and cuttings disposal,
(ii) produced water,
(iii) drainage water,
(iv) sewage systems,
(v)   atmospheric emissions,
(vi) noise.

Clearly all statutory requirements for emission monitoring must be met but considerable emphasis must also be placed on monitoring and recording procedures which provide:

(a) meaningful feedback to assess plant performance,
(b) input data for environmental monitoring (pollutant budgets).

EIA and environmental audit of present offshore oil production operations identify a range of potential impacts on the marine environment, mostly minor and local in nature (largely restricted to within the 500 metres safety zone around an installation). Two inputs present a more significant threat to the marine environment:

(i)   drilling cuttings disposal,
(ii)   produced water discharges.

Drilling Cuttings Disposal

There have been innumerable papers published relating to the disposal of drilling cuttings contaminated with drilling fluids. Although until the late 1970s petroleum products have been used for decades in the formulation of drilling fluids throughout the world, drilling muds have been primarily water-based (WBM). From the vast literature on US studies of water-based muds and their environmental impact (e.g. Ayers et al., 1980) little risk could be identified from the dumping of untreated cuttings and used fluids. Recent studies, however, also in the US, have indicated that the organic components in water-based muds can create considerable oxygen demand (BOD and COD) which could present organic enrichment and associated BOD problems in superficial sediments adjacent to the platform or in the water column if the discharge plume does not disperse rapidly, e.g. lying along a thermocline. Yet further US studies suggest that the practice of adding oil (diesel and other mineral oils) to water-based muds during drilling was more widespread and frequent than expected, and would enhance BOD problems (reviewed in Petrazzuolo et al., 1985).

The major environmental problem, however, has arisen with the large scale use of oil-based muds (OBM) in North Sea operations.

OBM Use

Several factors led to the rapid introduction of OBMs:

(i)   primary justification based on the:
  (a)   use of highly deviated drilling to deep North Sea formations with associated torque problems;
  (b)   water sensitivity of strata in North Sea formations, particularly shales, clays and salt zones;
(ii)   gap in UK oil pollution legislation resulting from interpretation of definition of oil, which excluded light petroleum derived oils (including lighter diesel oils) from prohibition of discharge to the marine environment (Side, 1986);
(iii)   drilling can usually be undertaken at faster speeds using turbo-drilling which needs the lubrication of OBMs.

With no restraints on the use and on-site disposal of oil-contaminated cuttings, such OBM use was extended from drilling only bottom hole sections of deviated wells to entire deviated holes and also vertical holes, including exploratory drilling. By 1982, fifty percent of wells drilled in the UK sector used OBMs (115 wells), resulting in the sea-bed dumping of at least 10,200 tonnes of oil (mainly diesel) as contamination on drilling cuttings. By 1984 this had risen to a total of 19,800 tonnes (Department of Energy, 1986).

Field (benthic) surveys close to platforms using OBMs in drilling operations have identified major impact on benthic macrofauna (e.g. Davies et al., 1984). The UK government attributed this impact largely to the toxic nature of diesel oil, particularly its aromatic content dominated by naphthalenes.

Reducing the environmental impact of OBM use could be achieved in several ways:

(i)   return to water-based muds;
(ii) reduce quantity of oil dumped by
      (a) restricting use of OBMs to bottom section of hole,
      (b) use of improved cuttings cleaning equipment, with reliable oil on
          cuttings monitoring procedures;
(iii) development of low-toxicity drilling fluids/muds.

Returning to water-based muds was considered impractical and industry was keen to use OBM for turbo drilling. UK experience suggested that the available status of cuttings cleaning technology was poor and major advances unlikely. With the UK firmly locked in to the importance of acute toxicity screening as the key to environmental protection from offshore chemical use (Voluntary Notification Scheme, see Blackman, 1984), the use of alternative mineral oil-based drilling fluids seemed the panacea. When new legislation was introduced the use of drilling fluids with low acute toxicity characteristics (96 hour LC_{50} brown shrimp test) was endorsed by Government, with no requirement for cuttings cleaning technology and no formal emission limit for oil on cuttings based on best practical means.

However, our audit procedure for drilling mud usage, and experience from related environmental/experimental studies, identify several major points as outlined below.

Nature of the Mud

1. There is a complete range of drilling mud combinations from
   water + barite (weighting agent); through a complex spectrum of water-
   based muds, including oil-in-water emulsion systems . . . to mineral

oil-based systems, initially diesel dominated (because it was available offshore and accepted for freeing stuck drill pipes, etc.) . . . to the range of new generation mineral oil-based systems.

2. Oil and/or water is only the base to a mud formulation which in turn must be capable of modification with additives during drilling.

   Many of these additives are much more toxic than the base oil and are usually carried in aromatic solvents e.g. oil-wetting agents can contain up to 30–60% v/v diesel and emulsifiers can contain up to 50% v/v high aromatic carriers.

3. The nature of the 'oil' contamination on cuttings will vary, differing considerably from the original drilling fluid. Changes can be due to such features as:

   (a) chemical changes downhole (temperature:pressure regime);
   (b) partition of specific compounds between fluids, and between fluids and particulates.

Even water-based muds of very low toxicity increase considerably in toxicity downhole (Neff *et al.*, 1981).

Concept of Toxicity and Drilling Mud Selection

Assessment of toxicity, particularly the applicability/transferability of data from laboratory to the field is notoriously difficult. It has cost millions of dollars in the drug industry and may well be being mis-used in the offshore situation. Particularly dangerous is the dependence on simple acute toxicity tests (such as the routine short-term LC_{50} test).

Dumping of cuttings contaminated with diesel-based muds result in several changes to the receiving marine environment:

(i) they will greatly alter the physical nature of sediments;
(ii) they will result in a very high organic enrichment, and as a consequence subsequent biodegradation will:
   (a) create high oxygen demand, raising position of sediment anoxic layer,
   (b) through further microbial action (SRB) result in build-up of sulphide levels (and reducing power);
(iii) several components will present both direct acute and chronic toxicity problems (notably aromatic hydrocarbons such as the naphthalenes).

By both direct and indirect routes diesel contamination presents a major threat to sediment fauna. Direct acute toxic properties will play a key role initially, with organic enrichment and consequent alteration of the sediment environment presenting the long term and dominant impact.

The extent of the impact will be primarily one of organic loading i.e. dumped cuttings volume and percent oil-on-cuttings, typical of organic waste inputs, (e.g. sewage, pulp waste). Acute toxicity may enhance the 'gross' impact but it may not be readily discernible.

Cuttings contaminated with the new 'low-toxicity' drilling muds present similar problems to the receiving environment:

(i) they will greatly alter the physical nature of the sediments (with some muds tested this could be a considerable problem),

(ii) organic enrichment of sediment — with consequent decrease of oxygen, increase in reducing power and formation of a toxic sulphide rich environment — presents the greatest threat, and

(iii) several additives to mud systems and downhole conditions could greatly alter the final toxicity of fluids on dumped cuttings.

With these new mineral oil-based systems (low aromatic base oils) the extent of impact will certainly be governed primarily by loading i.e. volume of cuttings times percent oil on cuttings. However the potential addition of toxic components to muds, some with less volatility than the naphthalenes of diesel could produce enhanced impact on benthic life.

It is unlikely therefore that the environmental impact (gram for gram) will be significantly different between diesel and the newer low aromatic oil-based mud systems and the way to minimize impact is to control volumes and oil levels of cuttings. Also, although short-term laboratory studies may suggest more ready/rapid biodegradation of the paraffinic oils, studies using laboratory continuous culture, sea-bed tank systems and direct oil field measurements suggest the rapid development of diesel degrading microbial populations. Related chemical and microbial studies demonstrate rapid loss of naphthalenes from cuttings piles on the sea-bed (Gillam et al., 1986). It is therefore likely that residual diesel contamination on cuttings may breakdown at similar rates to the new mineral oils, giving equivalent site recovery rates. Some evidence even suggests some native microbial populations favour the broader substrate spectrum provided by diesel-based muds (Gillam et al., 1986).

Recommended Operational Practice

Extensive diesel use may not be recommended on grounds of adverse effects on drilling personnel i.e. safety/health not marine environmental criteria.

Volume of discharged oil must be reduced (and more reliably monitored), by:

(i) reduction in amount of drilling with oil-based muds (only bottom sections of deviated wells);

(ii) reconsideration of other mud systems suitable for drilling deviated wells in water sensitive strata e.g. some of the oil-in-water emulsion mud systems;

(iii) improvement in control of oil on cuttings by solids control and/or cuttings cleaning systems.

There is potential to consider muds with lower oil levels and a recent Institute study certainly identified potential for improved cuttings cleaning systems (Davies, 1986). In fact the rush to the use of new mineral oil-based muds without the need for cuttings cleaning severely hit a developing UK capability in treatment technology.

Monitoring Environmental Impact

To date methods have centred on traditional monitoring of sediment chemistry and benthic macrofauna community structure.

Current findings confirm the major impact of organic enrichment, with 'traditional' dominance by opportunist fauna close to platforms (e.g. Matheson et al., 1986). Recently Kingston (1986) emphasized the high productivity of such benthic opportunists, apparently deriving food from the active microbial populations degrading the contaminating oils (diesel and mineral oils), e.g. vast polychaete populations flourishing on/in diesel contaminated cuttings with oil levels which would have killed the same animals in a laboratory LC_{50} test.

Kingston also claims that, given the available data, the only evidence of acute toxicity acting in addition to the organic enrichment effect comes from fields using the 'so-called' low-toxicity muds! Long term laboratory studies in Canada also suggest a build up of toxic effects in sediments (experimentally) contaminated with 'low toxicity' oiled cuttings (Barchard and Doe, 1984). It is interesting to quote from the Canadian work:

'we have identified that certain species show significant mortality after exposures of up to 32 days. It is noteworthy that if we had looked for only 96 hours we would have seen little or no effect'.

Experimental studies in Norway suggest that after 9 months on the sea-bed mineral oil contaminated cuttings increasing considerably in toxicity (Bakke et al., 1986).

Future monitoring should include rapid screening to assess the extent of impact from organic enrichment including in situ redox determination; supported by monitoring of recovery potential (microbial activity, in-fauna productivity); against a wider scale monitoring within the oilfield area to determine the extent of measurable impact. Recent evidence suggests that preoperational hydrocarbon levels at some new fields are higher than would have been predicted from the earliest North Sea studies. In the east of Shetland basin this background increase is in the range 2–3 times (over period 1978–1985). Gas chromatographic examination of aliphatic fractions suggest that although some of this background contamination is apparently drilling mud (diesel) derived there is also evidence of a much wider petroleum spectrum suggesting contamination from produced water and/or crude spillage sources. In some areas resolvable n-alkanes in the range nC_{12} to C_{25} suggest fresh inputs of diesel.

Environmental monitoring should be more closely linked to knowledge of drilling operations, particularly a more thorough and reliable monitoring of cuttings discharge. Also an ongoing mass-balance of mud use should be maintained. To be slightly controversial the present reported levels for oil input to the sea via cuttings dumping (which reached 19,800 tonnes in 1984) may be cause for concern but our attempts to 'audit' estimated sales/recovery of mud and declared disposal levels could not account for 35,100 tonnes in that same year. Where did it go, particularly the diesel-based fluid/mud?

Produced Water Discharge

Early in the development of North Sea oil reserves, produced water was identified as a major routine input of oil to the marine environment, which would increase steadily as the reservoir was depleted (CUEP, 1976).

Hydrocarbons were considered the major problem although, unlike the later response to cuttings disposal emphasis was on available technology for removing oil from effluent. Emission standards were based on best practical means, monitoring effluent oil levels as 'total oil' (largely free/dispersed oil) with the preferred method based on solvent extraction and infra-red measurement. No attempt was made to consider monitoring aromatic hydrocarbons, although audits had identified these as the hydrocarbons presenting greatest environmental risk (Johnston, 1980). It is interesting to note that now under the same Prevention of Oil Pollution Act (1971), control of oil on cuttings focuses on aromaticity and environmental risk.

Audit of a typical offshore operation identifies such factors as the following, which could influence produced water discharges.

1. Available technology (1970s) would have difficulty in maintaining low total-oil levels (below 40 ppm) as produced water volumes increased, and could be very sensitive to upset conditions.
2. Hydrocarbons in produced water which present the greatest risk to environment include aromatics (benzenes, naphthalenes), phenols and various other oxygenated compounds.

3. From studies worldwide, it has been observed that formation water can often have a relatively high BOD and COD, largely derived from such compounds as fatty acids.

4. Industry experience suggests that as a reservoir ages, increasing use will be made of chemicals, with many of these entering produced water, albeit some in modified form.

5. Formation water contains a wide range of metals particularly strontium and also enriches with metals from drilling operations (and chemicals from well stimulation).

6. Water injection practice often results in many downhole problems requiring chemical treatment (e.g. compatibility problems such as scale formation). This water, with its chemical loading will eventually break through into produced water.

In this paper, emphasis focuses on chemical usage offshore and their fate within production operations, particularly the consequences of injection water treatment and subsequent breakthrough. This area of chemical usage offshore is very complex and environmental implications are considerable and little understood — a key area for detailed audit procedures. How much of what goes in, where does it go, where might it appear and what is the environmental significance? The authors are convinced that such detailed chemical auditing is not only environmentally essential but the understanding gained will result in a reduction of process upset frequency and unnecessary chemical costs to the operator.

Crucial to the environmental audit of chemical use is an understanding of the different processes leading to chemicals entering effluent discharges. Offshore this is primarily the produced water discharge, but obviously includes pipeline export and subsequent water treatment at onshore terminals.

There has been considerable confusion relating to the possible breakthrough of injection water, often it seems reflecting a lack of understanding of oilfield systems. In 1979 it was considered (Department of Energy) that chemicals used in injection water would only reach the sea in small quantities from filter back washing, etc. (Read, 1980).

Government opinion as recently as 1985 (Bedborough, 1985) 'oxygen scavengers, scale inhibitors and corrosion inhibitors are used to treat injection water and will undergo chemical change before re-appearing (if at all) in the produced water . . .'.

Our audits have always assumed injection water breakthrough (e.g. Johnston, 1979), with most North Sea fields showing major breakthrough now. Our studies, and work in the US, have demonstrated carry-over of several injection water chemicals including biocides such as glutaraldehyde (Middleditch, 1984).

Figure 2 lists some of the major categories of chemicals used in oil and gas production offshore. In considering the potential fate and effects of

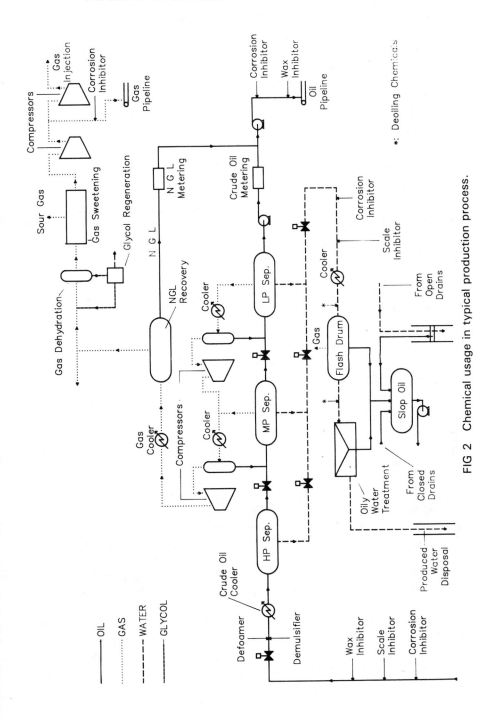

FIG 2   Chemical usage in typical production process.

these chemical categories it is crucial to emphasize that many are complex mixtures usually of crude grade chemicals (pot brew). Although data may be available, at least on the generic nature of their 'active' components, little information is given on additives, including solvents and dispersion aids. Frequently it is these secondary components that present major downstream problems.

Perhaps it would be appropriate to concentrate primarily on two categories of product to illustrate audit problems:

(i)   corrosion inhibitors
(ii) demulsifiers.

Corrosion Inhibitors

There is an extensive literature relating to the use of corrosion inhibitors in the oil industry (e.g. Bregman, 1963; Kelley, 1983; Goodson et al., 1985).

The most commonly used types are nitrogenous in nature, classified by Bregman (1963) as:

(i)   amides/imidazolines;
(ii)   salts of nitrogenous molecules with carboxylic acids;
(iii) nitrogen quaternaries;
(iv)   polyoxyalkylated amines, amides and imidazolines;
(v)   nitrogen heterocyclics.

Some less frequently used inhibitors have bases containing phosphorus, oxygen or sulphur in place of, or in addition to nitrogen.

If a corrosion inhibitor is to perform its primary role it must perform adequately at the site requiring corrosion protection, in an environment governed by such parameters as temperature, pressure and the total fluid regime. To establish such compatibility, particularly with the often complex mix of other chemicals in the system (e.g. demulsifiers, biocides, scale inhibitors, etc.), several modifying components may be required in the formulation e.g. antifoamers, demulsifiers or emulsifiers. Within an oil:water system the inhibitor must be transported to the site for protection then associate/bind with the metal surface requiring protection. To meet the varying needs products show a wide range of water:oil solubility and dispersibility.

The active components themselves can vary considerably in their solubility, dispersibility, with derivatives altering the characteristics of the parent compound. For example with imidazolines

$$R-C{=}N$$

(imidazoline ring structure with N–R^1, CH$_2$, CH$_2$)

changing R or R^1 can greatly alter the water (brine) and/or oil solubility of the inhibitor, thus (Goodson *et al.*, 1985):

inhibitor	brine solubility	oil (kerosine) solubility
1. naphthenic acid imidazoline	very slightly dispersible	soluble
2. fatty acid imidazoline	very slightly dispersible	soluble
3. imidazoline	soluble	insoluble
4. carboxylic acid salt of a fatty acid imidazoline	slightly dispersible	soluble

Inhibitor water solubility or dispersibility can also be enhanced by addition of solvents and surfactants. Imidazolines may also be quaternized. Similarly inhibitors which are water soluble can have their oil solubility enhanced with aromatic-based solvents.

The partitioning of corrosion inhibitors in water-hydrocarbon systems plays a key part in determining the eventual destiny of the inhibitor molecule (as does solubility and mode/point of injection). Obviously good understanding of inhibitor partition is essential for adequate corrosion protection, but it is also crucial in environmental audit — to help track the paths of particular chemicals in process systems and predict likely concentrations arriving in separation systems, and ultimately water discharges.

The partition of inhibitors between oil and water can be extremely variable, for instance in a study by Haslegrave (1985) partition coefficients ranged from 0.01 to 5.6, where

$$\text{Partition Coefficient (P.C.}_{w/o}) = \frac{C.I._w}{C.I._o}$$

He quotes an example of pipeline corrosion inhibitor treatment requiring 50 ppm C.I. in the water, using a water soluble inhibitor with a P.C.$_{w/o}$ = 5.

Pipeline fluids were 99.5% oil and 0.5% water. The required dosing of the total fluids was therefore only 10.2 ppm.

In many treatment operations dosing is at target strength, often with only rough data, if any, on partition coefficients. The consequence is a considerable excess of chemical in water, for treatment and discharge — also a considerable waste of money. Also when information is available it is usually on the main component, additives can have very different coefficients. In the above example dosing at 50 ppm would have given water concentrations of 245 ppm.

Some substances with very powerful attraction to water, e.g. the glycols, could reach very high concentrations in water and of course create considerable problems downstream. In general terms they would create high BOD in discharged water, and, in the case of glycols greatly alter the solubility/performance of many other chemicals (e.g. demulsifiers).

Corrosion inhibitor formulations can also contain several other treatment chemicals e.g. blends with biocides, scale inhibitors, etc.; all with different solubility/partition characteristics.

Demulsifiers

Both 'natural' petroleum components and upstream chemical treatment chemicals can have emulsifying properties, often resulting in serious treatment difficulties in separators and desalters. Emulsion breaking is primarily a physical process, usually aided by chemical treatment. As with all oilfield product categories there is a wide range of demulsifier formulations. Also, as with all products, they are not composed of individual chemicals with defined structures, so only an indication of their identification or functions (detail) can be given (also emphasized in a recent API study — Middleditch, 1984).

A group of commonly used demulsifiers is based on oxyalkylated phenol formaldehyde resins along with combination of some of the following additives and solvents:

polyglycols,
acylated polyglycols,
oxyalkylated alkanolamines,
other oxyalkylated phenols,
aryl sulphonates,
fatty acids (typically M.Wt. 800),
fatty alcohols (C_6–C_8),
aromatic hydrocarbons — xylene or a 'cut'
isopropanol/propanol,
methanol.

Oxyalkylation is employed to modify the solubility or dispersibility of the phenol formaldehyde resin, with ethoxylation or propoxylation being common. Solvents such as xylene also assist in solubilization.

A typical 'mix' commonly used is:

'oxyalkylated phenol formaldehyde resins plus aryl sulphonates and oxy-alkylated alkanolamines in aromatic hydrocarbons and isopropanol . . .'

To perform their de-emulsification role these products must function at the water-oil interface, certainly with some of their components entering the water phase. Some of the aromatic additives and solvents can reach high concentrations in the water phase e.g. desalter effluent, on occasions introducing higher levels of dissolved/dispersed aromatic hydrocarbons than in the original (pipeline) water. Downstream treatment, monitoring and final environmental risk can vary considerably, and suddenly, as rapid demulsifier dosing responses follow emulsion problems (e.g. in desalters).

With increasing water cuts, greater upstream use of other chemicals, including perhaps EOR chemicals, demulsifier use can only increase.

Present assessment of chemical use offshore is based on a voluntary scheme set up by the Department of Energy (Blackman, 1984). This scheme identifies substances whose use should be avoided (List 1) and those whose use could have concentrations in excess of 100 ppm in offshore discharges (List 2). (These categories are based on the Paris Convention).

However, central to the scheme is a classification of chemicals for offshore use, where suppliers are expected to provide data on product composition (physical and chemical) and an indication of its toxicity. Additional toxicity data may be required, based on a standardized 96 hour LC_{50} test using *Crangon crangon* (brown shrimp) and sometimes fish. A ranking (scale) 1–5 and 0 has been established indicating level of permitted use. Although the product classification has been of some success, albeit based on acute toxicity ranking usually unsupported by adequate chemical data, the failure has been at the next stage, when operators supply details on the use of specific chemicals particularly their likely fate/entry into discharges.

When refiners ask if chemicals are likely to get into the exported crude — how often is the answer no? Yet with an environmental enquiry the chemicals never seem to get into the produced water? Most of the product formulations have strongly surface active components, often the main chemical, so one can expect very complex partitioning patterns with many chemicals entering the water phase, often enriching above original dose levels. US studies have demonstrated the discharge of biocides such as glutaraldehyde and alkyldimethylbenzyl chloride (QUAT); and divers working near the Buccaneer Field experienced eye and skin irritations sufficient to interrupt their activities when acrolein was being used as a biocide (Middleditch, 1984).

As the larger fields age, we are already seeing the spectrum of chemical treatment increasing rapidly. A more formal auditing of chemical use and fate is essential. This could best be achieved by some general statutory obligation to provide evidence of an adequate environmental control strategy for offshore operations (Johnston and Side, 1985).

Within such a general control, specific attention could be given to the audit of chemical use and discharge, linked to a required monitoring of discharges and the receiving environment. As previously suggested (Johnston, 1984), with the volumes of produced waters discharged we need a more accurate understanding of expected dispersion patterns/rates.

It is also important that a better understanding and documentation of the chemical formulations of oilfield chemical products is obtained. As already indicated many formulations are extremely complex with secondary components (not documented) often producing the major downstream/ discharge treatment problems. Also, as products are often made under licence, with raw materials available, in some cases not every supplier knows the exact formulation — far less 'blenders' and ultimately users. Product quality assurance requirements are integral to the required audit procedure.

The Institute is currently developing an environmental audit procedure (microcomputer based) deriving mass balances for primary operations/ emissions:

I — Drilling Operations	cuttings and mud discharges;
II — Production Process	produced water discharges; atmospheric emissions (not discussed in this paper).

The drilling audit will address drilling procedures, well characteristics, mud characteristics (base fluid(s), standard mud additives and further mud modifying procedures offshore, with detailed attention on chemistry of all additives), solids control equipment and performance (for regular monitoring), any cuttings treatment and disposal systems (again with data on nature and levels of disposed cuttings and associated fluids/chemicals).

The audit of production operations will be based on a mass balance of the fluids involved, with detailed inputs of oil:gas:water production rates. Detailed audit of process water — nature/volumes of formation water, injection water characteristics, particularly added chemicals and history of injection water breakthrough. The detailed audit of all chemicals used will consider such factors as:

(i)   shore base — shipment — offshore use;
(ii)   detailed specifications of formulations*;
(iii) dosing — levels — positions frequency — monitoring checks.

Partition/solubility data (known/estimated) will be employed to:

(i)   predict distribution of specific chemicals throughout system;
(ii)   flag possible risk of interaction (downstream);
(iii) predict expected levels in discharged water.

REFERENCES

Ayers, R. C., Richards, N. L. and Gould, J. R. (Eds), (1980). *Proceedings of Symposium of Research on Environmental Fate and Effects of Drilling Fluids and Cuttings, 21–24 January 1980, Lake Buena Vista, Florida*, American Petroleum Institute, Washington DC.

Bakke T., Blackman, R. A. A., Hovde, H., Kjørsvik, E., Norland, S. Ormerod, K. and Østgaard, K. (1986). 'Drill cuttings on the sea bed — toxicity testing of cuttings before and after exposure on the sea floor for 9 months', in *Proceedings of Conference on Oil-Based Drilling Fluids — Cleaning and Environmental Effects of Oil Contaminated Drill Cuttings*, 24–26 February 1986, Trondheim, Norway, pp. 79–84, SFT/Statfjord Unit/Mobil Norway.

Barchard, W. and Doe, K. (1984). 'Preliminary results of bioassays of oily cuttings from the Alma F-67 exploration drilling programme — long term exposures', in *Oil-Based Mud Workshop. Environmental Protection Branch Technical Report No 2*, p. 58 (and pp. 20–34 in Appendix B), Canadian Oil and Gas Lands Administration, Ottawa.

Bedborough, D. R. (1985). 'Notification scheme for the selection of chemicals for use offshore', in *Chemicals in the Oil Industry Symposium*, 26–27 March 1985, University of Manchester, pp. 351–358, Royal Society of Chemistry, London.

Blackman, R. A. A. (1984). 'The UK notification scheme for the selection of chemicals for use offshore', in *Proceedings of a Symposium on Industrial Biocides, the Environment and Legislation*, 5 April 1984, pp. 31–41, Institute of Petroleum, London.

Bregman, J. I. (1963). *Corrosion Inhibitor*, Macmillan, New York.

CUEP (1976). *The separation of oil from water for North Sea oil operations*, a Report by the Central Unit on Environmental Pollution, Department of the Environment Pollution Paper No 6, HMSO, London.

Davies, J. M., Addy, J. M., Blackman, R. A., Blanchard, J. R., Ferbrache, J. E., Moore D. C., Sommerville, H. J., Whitehead, A. and Wilkinson, T. G. (1984). 'Environmental effects of the use of oil-based drilling muds in the North Sea', *Marine Pollution Bulletin*, **15** (10), 363–370

Davies, S. R. H. (1986). 'An assessment of the development of cuttings cleaning systems within the context of North Sea offshore drilling', in *Proceedings of Conference on Oil-Based Drilling Fluids — Cleaning and Environmental Effects of Oil Contaminated Drill Cuttings*, 24–26 February 1986, Trondheim, Norway, pp. 71–78, SFT/Statfjord Unit/Mobil Norway.

---

*More detailed product specification is essential preferably with help of suppliers/manufacturers, but screening may be necessary in QA back-up. As with other multi-operator studies we have been involved in a confidential form of data packaging/presentation is incorporated into the system under development.

Department of Energy (1986). *The Development of the oil and gas resources of the United Kingdom*, (The Brown Book), HMSO, London.

Gillam, A. H., O'Carrol, K. and Wardell, J. N. (1986). 'Biodegradation of oil adhering to drill cuttings', in *Proceedings of Conference on Oil-Based Drilling Fluids — Cleaning and Environmental Effects of Oil Contaminated Drill Cuttings*, 24–26 February 1986, Trondheim, Norway, pp. 123–136, SFT/Statfjord Unit/Mobil Norway.

Goodson, A. R., Kelley, J. A. and Jackson, J. E. (1985). 'A comparison of corrosion inhibitor performance tests for application in oil production systems', in *Chemicals in the Oil Industry Symposium*, 26–27 March 1985, University of Manchester, pp. 275–302, Royal Society of Chemistry, London.

Haselgrave, J. A. (1985). 'Partitioning of corrosion inhibitors in water/hydrocarbon systems', in *Chemicals in the Oil Industry Symposium*, 26–27 March 1985, University of Manchester, pp. 303–320, Royal Society of Chemistry, London.

Johnson, C. S. (1979). 'Environmental laws must be rooted in science not economic motives', *Offshore Engineer*, November, p. 64.

Johnston, C. S. (1980). 'Sources of hydrocarbons in the marine environment', in *Oily Water Discharges: Regulatory, Technical and Scientific Considerations* (Eds. C. S. Johnston and R. J. Morris), pp. 41–62, Applied Science, London.

Johnston, C. S. (1984). 'Fate of biocides in the environment', in *Proceedings of a Symposium on Industrial Biocides, the Environment and Legislation*, 5 April 1984, pp. 12–29, Institute of Petroleum, London.

Johnston, C. S. and Side, J. (1985). 'Safety and environmental aspects of offshore chemical usage — a review of national requirements and company policies', in *Chemicals in the Oil Industry Symposium*, 26–27 March 1985, University of Manchester, pp. 359–380, Royal Society of Chemistry, London.

Kelley, J. A. (1983). 'The chemistry of corrosion inhibitors used in oil production', in *Chemicals in the Oil Industry Symposium*, 22–23 March 1983, University of Manchester, pp. 150–158, Royal Society of Chemistry, London.

Kingston, P. F. (1986). 'Field effects of platform discharges on benthic macrofauna', *Royal Society Discussion Meeting on The Environmental Effects of North Sea Oil and Gas Developments*, 19–20 February 1986, London. (Proceedings in press).

Matheson, I., Kingston, P. F., Johnston, C. S. and Gibson, M. J., (1986). 'Statfjord field environmental study', in *Proceedings of Conference on Oil-Based Drilling Fluids — Cleaning and Environmental Effects of Oil Contaminated Drill Cuttings*, 24–26 February 1986, Trondheim, Norway, pp. 3–16, SFT/Statfjord Unit/Mobil Norway.

Middleditch, B. S. (1984). *Ecological effects of produced water discharge from offshore oil and gas platforms*, American Petroleum Institute, Washington DC.

Neff, J. M., Carr, R. S. and McCulloch, W. L. (1981). 'Acute toxicity of a used chrome lignosulfonate drilling mud to several species of marine invertebrate', *Marine Environmental Research*, **4**, 251–266.

Petrazzuolo, G., Michael, A. D., Menzie, C. A., Plugge, H. and Zimmerman, E. J. (1985). *Assessment of environmental fate and effects of discharges from offshore oil and gas operations*, EPA 440/4-85/002, Environmental Protection Agency, Washington DC.

Read, A. D. (1980). 'Personal Communication to Professor C. S. Johnston', *Institute of Offshore Engineering*, from Department of Energy, ref PET 215/1298/1.

Read, A. D. and Blackman, R. A. A. (1980). 'Oily water discharges from offshore North Sea installations: a perspective', *Marine Pollution Bulletin*, **11** (2), 44–47.

Side, J. (1986). 'Oil based muds: understanding the legislation', *Marine Pollution Bulletin*, **17** (3), 88–91.

BIODEGRADABILITY OF PRODUCED WATER EFFLUENTS

R. J. Watkinson and M. S. Holt

Shell Research Limited, Shell Research Centre, Sittingbourne, Kent, ME9 8AG

ABSTRACTS

The water co-produced with oil contains inorganic and organic components that are readily utilizable by microbial populations. In the marine environment, normally regarded as being nutritionally deficient, it can be shown that the major fraction of the total organic material is biodegraded.

INTRODUCTION

The operations involved in the winning of oil produce waste streams for discharge to the environment. The major waste stream, in terms of volume, during the production of oil is production water. The water is co-produced with the oil from the reservoir and separated by physical means before being discharged. The proportion of water to oil produced is not constant for the lifetime of the reservoir but increases from an almost zero water cut at the beginning of the reservoir production to perhaps 70–80% water cut or until it is no longer feasible to economically recover the oil. The water associated with the oil in the reservoir is sometimes referred to as formation water or connate water and makes up the major portion of production water. However, in fields where water flooding is practised then 'break-through' of injection water can occur and may make up to as much as 25% of the total water production. Beyond well-head, other processes may contribute to the water effluent. In certain offshore locations there are storage systems which rely upon seawater displacement to maintain the import/export of oil. This water is also subjected to oil-water separators before discharge and is usually kept separate but may become part of the production water train. These various sources of water that contribute to the overall 'produced water' create variations in concentrations of components from any one source. In addition to the oil and water, chemical additives may be present as oxygen scavengers, corrosion inhibitors and agents (demulsifiers) to aid the separation of oil and water.

The chemical composition of production water components will be varied, with the organic composition reflecting oil components, products of biogenesis, and production chemical additives. The fate of these components when discharged into the environment will depend upon the chemical and physico-chemical properties of the individual components, the properties of the receiving environment and the metabolic versatility of the biota in that environment.

The processes involved in the fate of the components will include evaporation, photo-oxidation and biodegradation along with the dispersive mechanisms such as adsorption, sinking and dilution.

The ecological effects of produced waters have been reviewed by Middleditch (Middleditch, 1984).

This paper is primarily concerned with the biodegradation of components of production waters.

Chemical Composition of Production Waters

Analysis of formation water is usually restricted to determining the concentration of inorganic species. Formation waters are generally saline but low in sulphate ions. Organic analyses for hydrocarbons have been carried out in fewer cases. The interest in the organic carbon present in formation waters has increased with the findings of high total organic carbon concentrations (> 1 g C/l) in some waters and the relevance of this to potential microbial activity within reservoirs. The soluble organic carbon in formation waters has special relevance to the potential souring activities of sulphate-reducing bacteria particularly where seawater injection systems will provide sulphate ions to the reservoir environment (Herbert et al., 1985).

Looking at the inorganic ion composition in formation waters from the view of supplying nutrients to micro-organisms one would conclude that there would be some limitations by the available phosphorus and possibly sulphur. Otherwise, with the relatively high concentrations of ammonium ions available (20–250 mg N/l) the waters would seem to provide a good marine medium for supporting microbial proliferation. Table 1 shows some inorganic constituents of production waters from a North Sea Field (Goodman and Troake 1984).

The Total Organic Carbon (TOC) of the production waters will consist of the suspended oil, the Dissolved Organic Carbon (DOC) and the possible carbon arising from production chemicals. In samples of production water taken world-wide the TOC values are in the range of 0–1500 mg/l.

Production waters, prior to discharge, are treated in gravity separators that will usually remove the suspended oil concentrations to below 25 ppm. However, such treatment processes will not remove any DOC that arises from dissolved hydrocarbon, phenolic compounds and simple fatty acids.

Table 1   Ionic composition of North Sea Water and Forties Field Production Water. (Goodman and Troake, 1984)

	Concentration (ppm)	
	Production Water	Sea Water
Anions		
Chloride	49550	19750
Sulphate	0	2650
Carbonate	0	0
Bicarbonate	425	140
Sulphide	0	0
Cations		
Sodium	32930	11200
Potassium	377	370
Calcium	2665	400
Magnesium	447	1400
Iron	3.3	0.5
Barium	218	0
Strontium	550	6
Boron	68	5.6

Table 2   Produced Water — Brent — typical composition

Component	mg C/l
Acetate	365
Propionate	63
Butyrate	20
Other acids	< 10
Phenols	< 2
Benzene	13
Toluene	8
Xylenes	3
Other volatile components	1–2
Inorganic carbon	41
Total	520

By far the major contributor to DOC in the production waters from production platforms are the simple fatty acids with acetate being the predominant component. Table 2 shows the composition of the fatty acids and soluble components found in North Sea production water (Somerville, 1984; Herbert *et al.*, 1985).

With this readily available carbon and nitrogen supply the expectation and observation is for the rapid biodegradation of these major components of production water.

Within the immediate vicinity of the discharge-point the environment will be confronted by an insoluble portion of the discharge (10–20 mg/l suspended oil) and a water soluble fraction. These two fractions of the discharge may well partition since the material buoyancy of the suspended oil should take it to the water surface whilst the soluble material will be dispersed in the water column. This partition of the components will lead to differences in local environments that each fraction is exposed to. It is generally accepted that surface films (water/air interface) carry greater microbial populations than the dispersed water column. However, it should be noted that suspended dispersed insoluble detritus will also have interfaces at which microbial communities concentrate.

Much research has concentrated on the biodegradation of oil films and this allows us to make clear statements on the biodegradability of the particulate oil discharged which is supported by data relating to specific oils and environments. However, some generalization can be made on the potential for biodegradation in the marine environment.

Biodegradation in Marine Environments

The marine environment is a heterogenous environment and can be simply divided for the purpose of these discussions into surface water film (seawater – air interface), main water column and sediment. The sediment may be sub-divided into the upper aerobic sediment and lower anaerobic zones. These are very general divisions and their extent and influence will vary according to local environmental conditions but they have important characteristics that inter-play with the microbial communities responsible for biodegradation.

In general terms the marine environment is generally regarded as being nutritionally deficient. The nutritional status of the major water column, particularly with respect to nitrogen and phosphorus is generally limiting the potential for microbial growth and activity. However, when nitrogen and phosphorus become available, either by upwelling from the sediments or some extraneous input, the response is usually the production of microbial blooms. In situations where little or no organic carbon is available the phytoplankton populations increase. When organic carbon is also available with the nitrogen and phosphorus supply then bacterial populations increase rapidly to consume the available nutrients. This rapid response represents the potential of an environment to degrade the organic inputs and fix available nutrients into biomass at the expense of oxidizing the organic material to CO_2 and water.

In attempting to define the response of a microbial population to production discharges the rate of degradation will be determined by the chemical composition and concentration of the inputs, the distribution of species within the indigenous population and the chemical and physico-chemical parameters of the receiving environment (indigenous nutrients, pH, temperature and dissolved oxygen). Degradation of the inputs will also take place by abiotic mechanisms including photo-oxidation, auto-oxidation and the discharged components will be further distributed by volatilization, solution and adsorption to particulate matter.

Bacteria and fungi can be regarded as being similar to the primary producers when high organic and nutrient concentrations become available. However, it is recognized that the true primary producers, phototrophic algae and phytoplankton may also carry out some modification and utilization of the production water components as well as higher members of the food chain. However, since their populations are seasonal and their rates of growth generally slower than bacteria, it is the bacteria which are generally considered as the most active agents in the mineralization of organic inputs.

The population of micro-organisms in marine water columns is usually of the order of 10^3–10^6 organisms/cm^3. This size of population is consistent with the measured total nitrogen values of 1–50 μg/l (ppb) which would support the growth of bacteria to the observed populations. The species diversity within the populations will reflect the organic inputs; there are wide variations in the genera that constitute the heterotrophic population. The hydrocarbonoclastic organisms generally represent less than 0.1% of the total population in non-polluted environments (Atlas, 1981). However, enrichment rapidly occurs in environments receiving hydrocarbons from spillages.

Micro-organisms themselves can exhibit some surface active properties and it is generally found that surface film populations are usually an order of magnitude greater than the water column population.

The most biologically active environment, in terms of populations or biological activity per unit volume, would appear to be the sediment. However, dependent upon the organic loading in the sediment, only the surface 2 cm is aerobic and below this the anaerobic bacteria predominate. The implications of this for sediment oil is that since true hydrocarbons are metabolized by mechanisms involving molecular oxygen their metabolism in deep sediments is limited by an oxygen requirement.

Thus it can be seen that the metabolic capability is ever present in the marine environment but its expression and capacity is limited by the nutritional status of that environment.

Biodegradation of Oil Films

The fate of oil and hydrocarbons in the marine environment has been extensively reviewed by Van der Linden, (1978) and more recently by Floodgate (1984). We have discussed in the previous section the dependence of microbial activity on nutritional status and the wide metabolic capability of the indigenous microbial flora. Oil hydrocarbons (alkanes and aromatics) represents one of the few chemical types whose metabolism demands the presence of molecular oxygen. The oxygen molecule is directly involved in initial attack on alkane and aromatic structures to produce alcohol, phenol and dihydroxy aromatic intermediates and metabolites, which may be further degraded to yield ultimately cell mass and carbon dioxide. Thus this demand for molecular oxygen precludes the biodegradation of such hydrocarbons in the anoxic zones of marine sediments and their degradation is most efficiently carried out in the water-surface film, water column and aerobic surface sediment. The intermediates, alcohols, phenols and acids which may also appear as minor constituents of oil, can undergo further transformation in anaerobic environments. Thus the limitation by oxygen affects only the initial aerobic microbial attack on the hydrocarbons.

Hydrocarbons have very limited water solubility and thus their availability to micro-organisms may limit their rate of degradation. The most soluble components are the monoaromatic hydrocarbons such as benzene, toluene and xylenes with naphthalenes being only sparingly soluble. Methane is soluble at mM concentrations at normal room temperature and pressure but as one ascends the homologous series of alkanes water solubility rapidly decreases such that octane can be considered as essentially insoluble. This means the micro-organisms take in their substrate directly from micro-organism/oil contact or in the form of an emulsion or micro-droplets (Watkinson, 1979) and thus the rate of hydrocarbon utilization can also be limited by the surface area of substrate available.

There are many general statements about the order of priority for the degradation of oil hydrocarbon components. They usually fall into the order alkanes > aromatics > iso-alkanes. However, it would obviously depend upon the metabolic capability of the community and certainly the rate of degradation of the more water soluble lower molecular weight species appears to be more rapid than the higher analogues. There seems little preference for aromatics or alkanes. However, there are obvious differences in the rates of degradation of alkanes and iso-alkanes such that the highly branched alkanes pristane and phytane are used as non-degradable internal markers. However, it is observed that even these bio-resistant molecules can disappear with time. Studies both in laboratory models and the field have demonstrated greater than 90% degradation of North Sea crudes both by biotic and abiotic mechanisms.

Biodegradation of Water Soluble Production Water Components

Experiments following oxygen comsumption in simple Closed Bottle experiments using production water diluted with seawater containing a

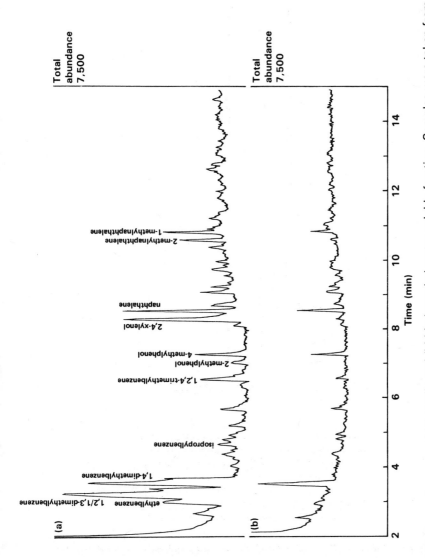

FIG 1   GC/MS chromatograms of (a) sterile and (b) biodegraded water soluble fraction. Samples were taken from continuous cultures with dilution rates of 0.021 hr^{-1}. The inoculated system contained a mixed marine microbial community isolated from the North Sea on water soluble oil fractions.

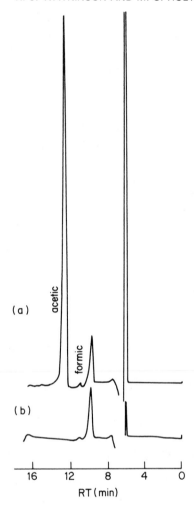

FIG 2   HPLC chromatograms of fatty acids of
production water (a) before and (b) after biodegradation with a mixed North Sea microbial
community. Samples were first treated with a
mixed bed ion exchange resin to remove inter-
fering ions and then separated on a Dionex
chromatograph using the HPICE AS-1 column.

mixed marine microbial population demonstrated a rapid oxygen consumption.
Such results indicated the components of production water to be very
readily biodegradable and the stoichiometry of the oxygen uptake for carbon
added indicated other oxidation processes to be taking place. This led to
the discovery of significant quantities of ammonium ions in the production

water which was being oxidized to nitrate by the microbial population. The rapid decrease in DOC in these experiments was found to be due to steam volatile materials from acidified production waters and identified as fatty acids with acetic acid as the major component.

Batch growth experiments, using production water as sole carbon and nutrient source, inoculated with mixed marine communities isolated from the North Sea, demonstrated rapid and extensive microbial growth. Gas chromatograms of the extracted dissolved hydrocarbons from samples of the culture broths taken at intervals demonstrated the removal of many of the hydrocarbon and phenolic compounds from the complex mixture (Figure 1).

New sample preparation and ion chromatography methods for the detection of fatty acids in seawater also allowed the monitoring of the rapid disappearance of the fatty acid components which were removed to the limits of detection (Figure 2). For North Sea production waters the disappearance of soluble hydrocarbons and fatty acids accounted for almost all the DOC removal by the microbial community.

Since simple fatty acids and particularly acetate, are involved in central metabolic pathways for living organisms it is no surprise that they are readily biodegraded.

In conclusion it has been demonstrated that the organic material discharged in production waters of offshore platforms provides a readily biodegradable carbon substrate for the indigenous microbial flora. These micro-organisms have the potential to rapidly mineralize this material to carbon dioxide and water as a result of their own proliferation.

REFERENCES

Atlas, R. M. (1981). 'Microbial degradation of petroleum hydrocarbons; an environmental perspective', *Microbiol. Rev.*, **45**, 180–209.

Floodgate, G. D. (1984). 'The fate of petroleum in marine ecosystems', *Petroleum Microbiology Ed. Atlas*, R. M. Macmillan, 355–397.

Goodman, K. S. and Troake, R. P. (1984). 'Environmental effects of production water discharges', *Petroleum Review*, August.

Herbert, B. N., Gilbert, P. D., Stockdale, H. and Watkinson, R. J. (1985). 'Factors controlling the activity of sulphate-reducing bacteria in reservoirs during water injection', *Conference proceedings. Offshore Europe '85, Aberdeen*. Society for Petroleum Engineers, SPE 13978.

Middleditch, B. (1984). 'Ecological effects of produced water effluents from offshore oil and gas platforms', *Ocean Management*, 191–316, Elsevier Science Pub.

Somerville, H. J. (1984). 'North Sea oil exploration and production. Interaction with the environment', *Instite. of Petroleum Ann. Conf., Aberdeen*, May 1984.

Van der Linden. (1978). 'Degradation of oil in the marine environment'. *Developments in biodegradation of hydrocarbons-1*. (ed.) Watkinson, R. J, Applied Science.

Watkinson, R. J. (1979). 'Interaction of micro-organisms with hydrocarbons', *Hydrocarbons in biotechnology*, Heyden Pub.

PROBLEMS ASSOCIATED WITH MARINE MICROBIAL FOULING

N. Gunn, D. C. Woods,
G. Blunn, R. L. Fletcher and E. B. G. Jones

*School of Biological Sciences,
Portsmouth Polytechnic, King Henry Building,
King Henry I Street, Portsmouth, Hants PO1 2DY*

GENERAL INTRODUCTION

All surfaces immersed in seawater undergo a series of discrete, sequential, chemical and biological changes. The earliest phase of this so-called biofilm development is dominated by the deposition of inorganic ions and high molecular weight polymeric compounds. These form a conditioning film approximately 50 Å in depth (Baier, 1973). By the use of scanning electron microscopy and infra-red spectophotometry it has been shown that continued exposure of the 'conditioned' surface results in bacterial colonization by both recruitment and growth, followed by the production of cellular exudations (Baier, 1984). The first colonizers are usually reported to be predominantly gram negative, rod-shaped bacteria such as species of the genera *Pseudomonas, Flavobacterium and Achromobacter*; later the generally larger, stalked, budding and filamentous bacteria usually move in and dominate the community e.g. species of the genera *Caulobacter, Hyphomicrobium* and *Saprospira* (Corpe, 1973; Dempsey, 1981). At this stage of fouling community development a large amount of detritus is captured by the presence of the bacterial adhesive. Secondary biofilm development is usually dominated by diatom colonization. The commonly recorded fouling genera include *Amphora, Achnanthes, Navicula* and *Cocconeis* (Daniel, Chamberlain and Jones, 1980; Robinson, Hall and Voltolina, 1985). Both the bacteria and diatoms tend to dominate the cellular fraction of any biofilm; these cells and their mucilaginous secretions constitute the majority of the biofilm organic content e.g. Caron and Seiburth (1981) concluded that diatoms make up 99.9% of adhered organic material. There are, however, a number of other taxa of microfouling organisms often found as constituents of biofilms; for example the fungal genera *Aspergillus* and *Penicillium* (O'Neill and Wilcox, 1971), various blue-green algae (Cyanophyceae) (Terry and Edyvean, 1981; McDonnell, Mulder, Hodgekiss

175

and Boney, 1984) and, amoeba, choanoflagellates, ciliates and protozoa have also been described (Caron and Sieburth, 1981; Brown and Jones, 1983).

On many surfaces these microbiological films are then settled upon by the spores and larval stages of macrofouling organisms. Indeed there is some evidence that the presence of a fouling biofilm positively promotes this settlement process (Barnes, 1970; Young and Mitchell, 1973; Mitchell and Kirchman, 1984; Huang and Boney, 1983; Thomas and Allsopp, 1983). Common fouling algal genera include *Enteromorpha, Ectocarpus and Polysiphonia* (Fletcher and Chamberlain, 1975; Fletcher, 1980) whilst fouling animals include *Balanus, Mytilus, Pomatoceros, Styela and Bugula* (Southward and Crisp, 1963; Ryland, 1965; Nelson-Smith, 1967; Millar, 1968; Leitch, 1980).

Traditionally most concern about the detrimental effects of fouling e.g. drag effect on ships' hulls, increased loading of platform supports has been associated wih the macrofouling communities. However, more recently increasing attention has been focused on the problems associated with the early formed microbial communities and the requirements for control measures. For example, although macrofouling on ships has been markedly reduced by the use of organotin containing ablative paints, slime films comprising bacteria, diatoms and blue-green algae are usually present (Bishop, 1969; Bishop, Marson and Silva, 1972; Bishop, Silva and Silva, 1974; Evans, 1981; Loeb and Smith, 1981). These not only probably effect the efficiency/longevity of the antifouling coatings, but they can, by themselves, considerably increase the drag on the ship's hull resulting in increased and expensive fuel consumption to maintain cruising speeds (Clitheroe, 1983). Microbial communities of bacteria, protozoa and ciliates also build up as slime on the darkened insides of condenser tubes, markedly reducing their heat transfer efficiency (Garey, Jorden, Aitken, Burton and Gray, 1980; Adamson, Liberator & Taylor, 1984; Somerscales and Knudsen, 1981), whilst bacteria, are known to be involved in the microbiological corrosion of metal surfaces (Gerchakov and Udey, 1984; Edyvean and Terry, 1983). It is with these microbiological problems, of particular importance to the offshore industry, that the School of Biological Sciences at Portsmouth Polytechnic is currently involved, and the present paper outlines some of the general findings of this research. In particular three separate aspects are considered: a) Microbial fouling of condenser systems. b) Microbial fouling of ships. c) Microbial fouling of metal surfaces.

MICROBIAL FOULING OF CONDENSER SYSTEMS

Introduction

A problem particularly applicable to the offshore industry is the fouling of heat exchange systems by marine growths. Seawater is constantly being used for cooling purposes either for 'domestic' air conditioning or for steam turbine condenser systems. The condensation of large amounts of steam is extremely difficult as a surface must be maintained at a sufficiently low temperature to allow the energy to be released from the steam, to form water. A wide variety of condensation equipment has been designed and tested all of which are based on the principle of presenting a continuously cold surface for condensation. The method most frequently used for power generation involves once-through tubular condensers $\simeq 10$ meters in length and 2 cm in diameter. These are packed together in bundles, up to 10000 in number, and seawater is passed through them at speeds of about $2\,m.s^{-1}$ which gives a suitable heat pickup whilst maintaining the energy required to pump the coolant at a suitable level.

For many years the materials used for condenser tubes have been either copper or its alloys. However, these are subject to corrosion problems so new alternative materials are being investigated. Out of a number of corrosion-resistant and inert materials examined titanium has proved the most successful. In addition to its anti-corrosion properties it has a high strength to weight ratio, which allows the tubes to be manufactured with much thinner walls (so improving the heat transfer across the tube wall) but without loss of strength.

Apart from being expensive the major disadvantage of titanium and many of the other high corrosion resistance metals is their well documented vulnerability to settlement and growth by a wide variety of marine fouling organisms (Marshall, Stout and Mitchell, 1971; Corpe and Winters, 1972; Loeb and Neihof, 1975; DePalma, Goupers and Akers, 1979; Characklis, 1981). These growths cause three man problems; a) enhanced corrosion (if a corrosive material is being used) b) increased fluid frictional resistance, giving rise to an increased pressure differential along the tube and c) increased heat transfer resistance. These fouling related problems have received particular attention over the last fifteen years (Corpe, 1970a; Corpe, 1970b; Marshall, Stout and Mitchell 1971a; Marshall et al., 1971b; Corpe, 1973; Characklis, 1973a; Characklis, 1973b; Loeb and Neihof, 1975; Norrman, Characklis and Bryers, 1977; Bryers and Characklis, 1979; Characklis, 1979; Nimmons, 1979; Characklis, 1981).

The effects of microbial fouling of titanium condenser tubes in comparison with cupro-nickel alloys will be discussed in this paper.

Materials and Methods

Titanium and 90:10 cupro-nickel tubes were exposed for different periods to seawater under a variety of conditions. At the end of the exposure periods the tubes were tested for heat transfer resistance and pressure differentials.

Table 1   Temperature differentials and percentage difference over controls. Titanium laboratory pumped samples

Flow M/S⁻¹	Exposure period												
	1wk		3wks		4wks		7wks		8wks		9wks		22wks	
	Δt	%	Δt	%	Δt	%	Δt	%	Δt	%	Δt	%	Δt	%
0.5	7.76	−1.9	6.50	−1.6	7.73	0.4	7.43	−4.3	7.49	−3.5	7.88	1.5	7.95	2.4
1.0	6.81	−1.9	5.75	−17.2	6.79	−2.2	6.38	−8.1	6.65	−4.2	7.02	1.1	7.20	3.7
1.5	5.30	−4.6	5.19	−6.5	5.25	−5.5	5.18	−4.7	5.24	−7.6	5.38	3.1	5.54	−0.2
2.0	4.51	−3.0	4.32	−7.1	4.41	−5.2	4.37	−6.0	4.3	−7.6	4.45	−4.3	3.55	−2.2
2.5	3.88	−7.3	3.72	−6.7	3.75	−6.0	3.70	−7.3	3.67	−8.0	3.81	−4.5	3.88	−2.7
3.0	3.40	−1.7	3.29	−4.9	3.23	−6.6	3.28	−5.2	3.27	−5.5	3.40	−1.7	3.37	−2.6
4.0	2.78	0.4	2.67	−3.6	2.61	−5.8	2.67	−3.6	2.66	−3.9	2.70	−2.5	2.72	−1.81
5.0	2.30	−0.4	2.23	−3.5	2.19	−5.2	2.23	−3.4	2.23	−3.4	2.33	−0.8	2.28	−1.3

Table 2 Pressure differentials and percentage difference over controls. Titanium laboratory pumped samples

Flow M/S⁻¹	1wk		3wks		4wks		7wks		8wks		9wks		22wks	
	ΔP	%	ΔP	%	ΔP	%	ΔP	%	ΔP	%	ΔP	%	ΔP	%
0.5	65	4.8	105	69	72	16	99	59	98	58	81	31	93	50
1.0	103	0.9	223	118	142	39	212	107	156	53	166	62	213	108
1.5	220	2.3	297	38	249	16	313	45	345	60	324	51	371	73
2.0	350	−.16	420	18	389	9.2	442	24	515	44	487	37	548	53
2.5	518	−2.1	587	11	566	7.0	614	16	677	28	678	28	734	38
3.0	734	−1.0	791	6.7	769	3.7	811	9.4	825	11	831	12	995	34
4.0	1222	−3.4	1285	1.5	1285	1.5	1329	5.0	1348	6.5	1424	12.5	1575	24
5.0	1852	−1.5	1928	2.6	1890	0.6	1984	5.5	1928	2.6	1991	5.9	2268	20

ΔP expressed in mmHg

Table 3 Temperature differentials and percentage difference over controls. Titanium Harbour pumped samples

Flow M/S⁻¹	Exposure period											
	3mns		4mns		5mns		6mns		9mns		12mns	
	Δt	%	Δt	%	Δt	%	Δt	%	Δt	%	Δt	%
0.5	8.40	8.2	8.20	5.6	8.00	3.2	8.36	7.7	8.36	12.8	8.31	7.0
1.0	7.55	8.7	7.50	8.0	7.20	3.7	7.28	4.9	7.65	10.2	7.37	6.2
1.5	5.90	6.3	5.70	2.7	5.59	0.7	5.74	3.4	5.80	4.5	5.60	0.9
2.0	4.86	4.5	4.70	1.0	4.56	−1.9	4.69	0.9	4.71	1.2	4.56	−0.2
2.5	4.15	4.0	3.90	−2.3	3.90	−2.3	3.99	0.0	4.02	0.8	3.91	−0.02
3.0	3.57	3.1	3.40	−1.7	3.40	−1.7	3.49	0.9	3.50	1.1	3.5	1.1
4.0	2.82	1.8	2.80	1.1	2.71	−2.2	2.76	−0.4	2.79	0.7	2.72	−1.8
5.0	2.35	1.73	2.30	−0.4	2.30	−0.4	2.29	−0.7	2.31	0.0	2.25	−2.6

Table 4 Pressure differentials and percentage difference over controls. Titanium Harbour pumped samples

Flow M/S^{-1}	Exposure period											
	3mns		4mns		5mns		6mns		9mns		12mns	
	ΔP	%	ΔP	%	ΔP	%	ΔP	%	ΔP	%	ΔP	%
0.5	168	170	176	174	280	351	124	100	239	285	262	322
1.0	273	167	300	194	797	681	262	156	408	300	503	393
1.5	534	148	594	176	1104	413	456	112	782	263	957	345
2.0	752	111	954	167	1560	338	688	93	974	173	1424	300
2.5	1027	94	1335	152	1903	259	903	70	1278	141	1916	262
3.0	1566	111	1713	131	2407	224	1249	68	1776	139	2533	241
4.0	2925	131	2961	134	3062	142	1916	51	3011	138	4982	293
5.0	3603	91	3503	86	4133	119	2790	48	4176	122	5657	201

ΔP expressed in mmHg

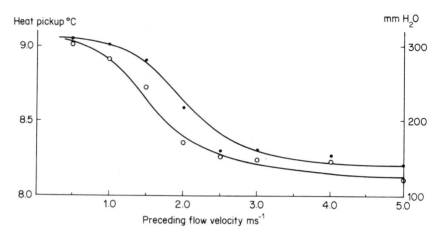

FIG 1   Relationship between heat pickup and back pressure at 0.5 m.s⁻¹.
● = temperature. ○ = back pressure mm H_2O.

The exposure regimes were as follows: a) Tubes were suspended from the Polytechnic's test raft in Langstone Harbour, Hampshire. This allowed slow passage of water through the tube permitting settlement of micro and macro-organisms. b) Tubes were exposed to atmospheric conditions outside the marine laboratory and seawater was pumped through them on a continuous basis at speeds of 1.5–2.0 m.s⁻¹. This was designed to imitate flow conditions in the condenser, and c) Tubes were exposed within the laboratory and had water pumped continuously through them at 1.5–2.0 m.s⁻¹. The water temperature was controlled at 15°C±1°C.

After the required exposure periods (Tables 1–4) the tubes were installed in a single tube model condenser. Steam was provided at 63.8°C and 27" Hg vacuum, figures comparable to working condensers. Cooling water was pumped through the centrally situated length of condenser tube at a controlled velocity, and at an inlet temperature of 20°C. When the parameters of inlet temperature, flow velocity, condenser temperature and vacuum were constant the outlet temperature was measured. This procedure was repeated at eight flow velocities, 0.5, 1.0, 1.5, 2.0, 2.5, 3.0, 4.0 and 5.0 m.s⁻¹ and the mean of three outlet temperatures for each velocity taken. The pressure differentials were also recorded on a water manometer for each velocity.

The procedure used to estimate the relationship between the two parameters was similar to that previously described, but with the heat pickup and pressure differentials only being measured at 0.5 m.s⁻¹. The flow was then increased to 1.0 m.s⁻¹ for 30 minutes, whilst still maintaining constant inlet temperature and condenser temperature and vacuum. The flow was then returned to 0.5 m.s⁻¹ and the outlet

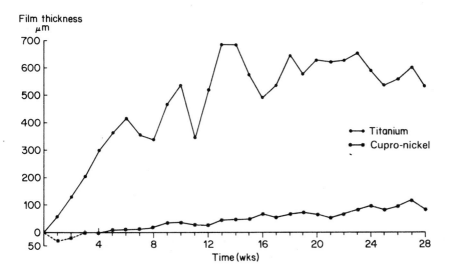

FIG 2   Development of wet film thickness with time. ● = Titanium, * = Cupro-nickel.

temperature and pressure differential recorded. This procedure was repeated with each of the following velocities 1.0, 1.5, 2.0, 2.5, 3.0, 4.0 and 5.0 m.s^{-1}. The results are presented in Figure 1. Estimates by weight of film thickness were made and these are also presented.

Results and Discussion

The results show that there is a dramatic, but expected, difference in fouling and performance between the inert titanium tubes and the toxic cupro-nickel tubes. Both the harbour fouled and 'pumped' titanium tubes were rapidly fouled whilst the cupro-nickel tubes supported little growth (Figure 2). The film within the titanium tubes increased rapidly in the first two weeks and then showed a steady increase up to 6 weeks. There was then a slight decrease in film thickness, followed by a very rapid and dramatic increase. This pattern was repeated throughout the trial with the fluctuations gradually flattening out, to give a wet film thickness of approximately 600 μm. The cupro-nickel tubes, however, initially showed a negative film thickness which was then followed by a very gradual and small increase. A negative film thickness was recorded because the measurements are calculated from weight variations of the tubes, the weight losses probably being due to corrosion.

The oscillation observed in film thickness in the titanium tube has been seen by other authors (Characklis, 1981) and is due to the sloughing off and redevelopment of the fouling layer. This occurs when the film reaches

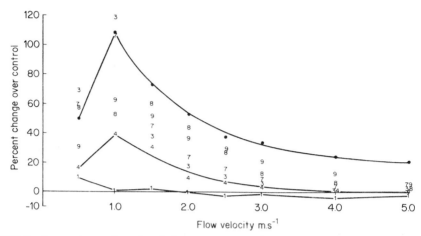

FIG 3 Percentage alteration in back pressure of pumped samples over controls. Sample N° 1–9 = wks, ● = 22 wks.

a critical film thickness of 350–400 μm and the strength of the bond between the base metal and the fouling material is not strong enough to withstand the shear force applied by the water movement. The critical film thickness gradually increases as the film becomes more complex and with greater surface/film contact.

The results of the pressure differentials (Figures 3 and 4) for the pumped and harbour fouled tubes show an initial rise followed by a steady and gradual decline. Although the pattern was similar the level at which this occurred was significantly different. The harbour fouled tubes showed an increase of 100% to 700% whilst the pumped samples reached a maximum of 120%. This is probably because macrofouling species readily settled wtihin the harbour fouled tubes, creating extreme turbulence, whilst within the pumped samples no macrofouling occurred which allowed laminar flow at the lower water velocities.

Conclusions

As expected titanium rapidly fouls with a wide variety of organisms developing on the surface of the tubing. The extent of the fouling is significant with a 700% increase in pressure differentials over the controls recorded within 5 months of exposure. This figure is tempered by the fact that the tubes were exposed to slow moving coastal surface waters whilst an in-service condenser would have a much faster flow rate. However, the conditions experienced by naval and civilian vessels in port are not dissimilar to those in this experiment, with slow moving surface water passing through condensers running only at low power. It is under these conditions

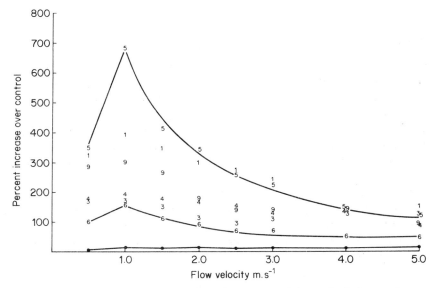

FIG 4 Percentage increase in back pressure of harbour fouled samples over controls. Sample N° 3-9 = wks exposure, sample N°1 = 1 year exposure, ● = Cupro-nickel sample.

that settlement of the micro-organisms occurs. We have shown that once material has settled it is particularly difficult to remove, at normal service operational velocities ($\simeq 2\,m.s^{-1}$ flow velocity). Furthermore, in Figure 1, it can be seen that an occasional increase in flow velocity to the region of $5\,m.s^{-1}$, which has been advocated as a cleaning technique (Conn, Rice and Hagal, 1977) is not sufficient for total removal of fouling films.

Our results show that titanium will suffer a detrimental loss in efficiency within 6 months of water passage at operational working velocities. It is not possible to categorically state, at this stage, that this loss in efficiency would occur under working conditions as the parameters of increased wall temperature and different microbial concentrations may effect the results. In this connection our preliminary studies appear to suggest that increased tube wall temperature has little effect on the development of the film and may in fact serve to enhance growth.

In contrast to this, cupro-nickel tubes do not suffer from severe fouling problems. However, it is not true to state that they do not foul, but that the material appears to be tenuously attached and easily removed by a flow increase to $5\,m.s^{-1}$. This is due to the unstable nature of the surface and the rapid corrosion which occurs (Blunn and Jones, 1985).

MICROBIAL FOULING OF SHIPS

Introduction

It has long been established that the build up of fouling organisms on ships' hulls incurs a financial penalty for the owners or operators (Dempsey, 1981; Christie, 1982; Fischer, Castelli, Rodgers and Bleile, 1984; Preiser, Ticker and Bohlander, 1984). Increased 'wave forming' and 'skin friction' resistance caused by fouling results in either loss of speed or greater fuel consumption. Additional financial loss can be incurred through mechanical breakdown of protective antifouling paint coatings by these fouling organisms (Moss and Woodhead, 1970).

Initial colonization of toxic antifouling paints is predominantly by bacteria and procumbent diatoms (*Amphora, Cocconeis, Navicula*) which form a characteristic 'slime film'. The inclusion of other groups of fouling organisms, especially reduced macro-algal forms (*Ectocarpus, Enteromorpha*) further facilitates colonization by epiphytic genera such as *Licmophora*. These epiphytic diatoms are also commonly attached to colonial diatoms such as tube forming species of the genus *Navicula*. It is evident that fouling films are a complex combination of 'direct' and 'indirect' attached fouling organisms. Due to the advances made in antifouling technology most animal and much macro-algal fouling is effectively controlled; it is therefore the occurrence of microfouling films which represents the greatest problem, of which diatoms generally form the largest entity (Caron and Sieburth, 1981). In the present study samples of microfouling organisms were collected from various sites on a range of in-service vessels in order to determine both biofilm structure and species composition.

Materials and Methods

Samples of microfouling organisms were collected from 15 in-service vessels between October, 1984 and March, 1986. These samples were still wet at collection and were preserved in formal saline. Laboratory studies involved the acid cleaning of the diatom frustules to facilitate accurate species identification and the use of a 'Light section measuring microscope' to produce comparative profiles of both paint and biofilm surfaces in addition to quantifying biofilm thickness. The latter measurements required intact fouling biofilms on fragments of antifouling paint.

Results and Discussion

Of the large number of diatoms described, including those occurring around the British Isles (Hendey, 1974), relatively few species can be termed

Table 5   Diatom composition of microfouling samples from a Mediterranean-based vessel. Abundance estimate scale: D = dominant (greater than 50%), S.D. = subdominant (25-50%), C = common (10-25%)

Sample site	No. of diatom genera/species principal fouling species	
1. Forward, 150 cm below waterline	17 genera: 30 species *Navicula ramosissima var. mollis*	D.
2. Forward, 300 cm below waterline	17 genera: 32 species *Amphora coffeaeformis var. perpusilla* *Navicula corymbosa*	D. C.
3. Forward/midship, 360–450 cm below waterline	14 genera: 25 species *A. coffeaeformis var. perpusilla* *N. corymbosa*	S.D C.
4. Midship, waterline	16 genera: 31 species *A coffeaeformis var. perpusilla* *N. ramosissima var. mollis*	S.D S.D.
5. Midship, 150 cm	16 genera: 29 species *A. coffeaeformis var. perpusilla* *N. ramosissima var. mollis*	D. C.
6. Aft/midship, 600 cm	12 genera: 27 species *A. coffeaeformis var. perpusilla*	D.
7. Aft, 600 cm	6 genera: 10 species *A. coffeaeformis var. perpusilla* *Navicula sp. (florinae ?)* *N. corymbosa*	S.D. C. C.

'successful foulers', namely those that can exhibit a range of morphologies and attachment mechanisms. The most common and cosmopolitan genus observed in the present study was *Amphora* (Table 5) of which *Amphora coffeaeformis var. perpusilla* appears to be the most prevalent species. This is a small species with the cells commonly closely adpressed to the surface and so offering a low profile to fluid shear forces. They often form fine slime films attached directly to antifouling paint surfaces and are noted for their resistance to cuprous oxide, a common antifouling toxin (Daniel and Chamberlain, 1981). A similar growth form and attachment mechanism was observed for the genus *Cocconeis*. This genus was observed quite commonly on a range of antifouling paints; although a number of species were identified, e.g. *C. scutellum, C. peltiodes* and *C. pellucida*, no single species was particularly prevalent. Also observed sporadically on ships was the stalk forming genus *Achnanthes*, members of which have previously been reported to dominate antifouling paints (Callow, Wood and Christie, 1976); species identified include *A. longipes, A. subsessilis* and *A. parvula*.

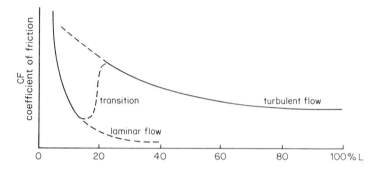

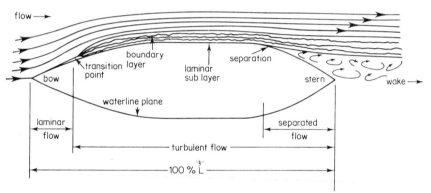

FIG 5   Transitional water flow about a ship's hull (after Gillmer, 1970).

A single cell attaches by producing a mucilage pad followed by the development of a stalk which elevates it away from the toxic surface. Members of this genus were observed on ships, either as 'direct' fouling organisms attached to the paint or as epiphytes. Another common genus was *Licmophora* which forms stalked and branched colonies. Studies of fouling samples indicates that this genus occurs mainly as an epiphyte on reduced macro-algal species, rather than as a direct colonizer of paint surfaces. Finally a number of *Navicula* species commonly occur on ships' hulls, the most abundant being *Navicula corymbosa*. Many members of this genus form colonies consisting of cells contained within mucilagenous tubes. The colony as a whole has an erect, tuft-like morphology and was often subject to colonization by other epiphytic diatoms, resulting in an increase in biofilm thickness and complexity.

The species composition of microfouling films on in-service shipping is subject to a number of selective influences. For example, vertical variations in the flora have been observed on most vessels which are probably associated with the degree of light penetration, a previously well-

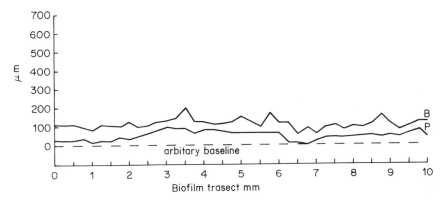

FIG 6   M.V. Concord, Bow, Keel, Transect taken 5.6 m below waterline. (B = Biofilm surface profile; P = Point surface profile).

documented phenomenon (Bacon and Taylor, 1976; Pyne, Fletcher and Jones, 1986, in press). On a ship's hull, however, especially at the waterline region other factors such as increased oxygenation and wave action will probably effect the floristic composition. For example, identification of fouling diatoms from the M.V. Brading showed the greatest species diversity to occur below the water line; samples between 50 cm and 120 cm below the water line contained an average of 28 species whilst water line samples contained only an average of 17 species.

The type of anti-fouling paint employed and the routes followed by the vessels have also been shown to influence species composition. Generally diatom fouling has been found to be more extensive on vessels operating in a limited area e.g. Solent and Channel ferries, than on vessels operating over global routes where fouling organisms would be subject to greater variations in temperature, salinity and nutrient availability. Samples from the submerged hulls of Solent-based vessels such as the M.V. Brading and the M.V. St. Catherine contained an average of 27 diatoms species, while ocean-going vessels (M.V. Concord and M.V. Rangelock) exhibited an average of only 12 species.

Horizontal variations in species composition frequently occur which are probably the result of changes in the hydrodynamic forces applied along a hull. Flow in the bow region of a vessel tends to be laminar, the extent of this is dependent on the shape of the hull. This then passes through a transition stage and becomes turbulent (Figure 5). Turbulent flow produces greater shear stress against a surface and consequently against any fouling organisms on that surface (Fowler and McKay, 1979). Flow in the stern region is 'separated' as a result of the vessel's passage and the action of the propeller increases turbulence in this region (Gilmer, 1970).

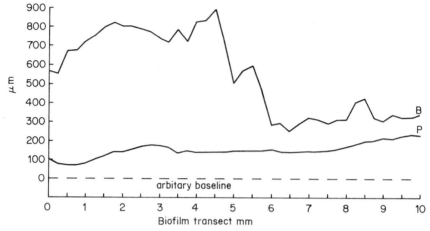

FIG 7   M.V. Concord. Midships, Transect taken 2.7 m below waterline.

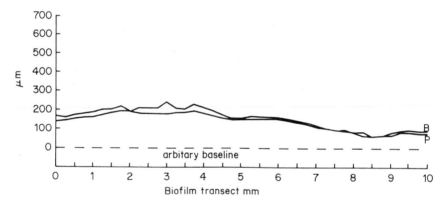

FIG 8   M.V. Concord, Aft. Transect taken 3.3 m below waterline.

The results (Table 6, Figures 6, 7 and 8) indicate that these variations in hydrodynamic forces on the hull probably cause local differences in both species composition and biofilm thickness. This was particularly prominent on the M.V. Brading, a Solent-based vessel which had a narrow bow and so exhibited the classic 'Bernuolli' flow pattern shown in Figure 5, with a large area of the bow subject to laminar flow. The bow region was extensively fouled by reduced macro-algae and epiphytic *Licmophora* species (Table 6). Towards the stern these were gradually replaced by much thinner, smoother slime films dominated by *Amphora coffeaeformis var. perpusilla* (Table 6).

The profiles shown in Figures 6, 7 and 8 are of microfouling films sampled from a bulk carrier, the M.V. Concord, after 15 months service in the

Table 6   Fouling diatoms identified from M.V. Brading. Samples taken from 120 cm below the waterline

Forward	Midship	Aft
Achnanthes subsessilis	Achnanthes longipes	Achnanthes subsessilis
Actinoptychus senarius	A. subsessilis	Amphora sp. (exigua?)
Amphora coffeaeformis var. perpusilla	Actinoptychus senarius	A. coffeaeformis var. coffeaeformis
A. turgida	Amphora coffeaeformi var. perpusilla S.D.	A. coffeaeformis var. perpusilla S.D.
A veneta	A. veneta	A. salina
Cocconeis sp.	Biddulphia aurita	Biddulphia aurita
C. peltiodes	Caloneis liber	Cocconeis costata
C. scutelum	Cocconei peltiodes	C. disculus
Diatoma vulgare	C. scutellum	C. peltiodes
Fragilaria sp.	Diatoma vulgare	C. scuttelum
Grammatophora marina	Diploneis didyma	Diatoma vulgare
Licmophora anglica S.D.	Fragilaria sp.	Fragilaria sp.
L. communis	F. harrisonii	Grammatophora marina
Navicula sp.	Grammatophora marina	Licmophora anglica
N. arenaria	Hantzschia sp.	L. communis
N. avenacea	Licmophora anglica	Navicula sp.
N. corymbosa	Melosira westii	N. corymbosa
N. cincta	Navicula sp. (× 3)	N. cryptocephala var. veneta
N. crucigera	N. calida	N. digito-radiata
N. grevilleana	N. corymbosa	N. pseudocomioides
N. ramosissima var. ramosissima	N. cryptocephala var. veneta	N. ramosissima var. mucosa
Nitzschia sp.	N. inflexa	Nitzschia sp.
Synedra fasciculata	N. ramosissima var. ramosissima	N. apiculata
S. gaillonii	Nitzschia sp.	Opephora marina
	N. dissipata	Rhiocosphenia curvata
	Raphoneis amphiceros	Stauroneis decipiens
	Stauroneis decipiens	Synedra fasciculata
	Synedra fasciculata	Thalassiosira sp.
	Thalassiosira sp.	

eastern Atlantic. This was a bulbous-bowed vessel and did not have a region of laminar flow at the bow. As a result the pattern of fouling on the hull was different to that of the M.V. Brading. The relatively thin fouling films measured in the bow region of the M.V. Concord were due to the high shear stress created by the forward movement of the vessel, and at the stern, they were due to the separated flow caused by the ship's passage. Hydrodynamic conditions in the midship region were most conducive to biofilm development and as a result the thickest fouling films were found here. Statistical studies using 'analysis of variance' show there to be a 15–30-fold increase in surface roughness due to fouling at the midship site and a 1–3-fold increase at the bow and stern sites. As a result the midship region would have a much greater influence on the frictional resistance of the vessel and thus on fuel consumption.

Conclusions

A number of conclusions can be drawn from the present survey of micro-fouling of 'in-service' shipping: a) Microfouling has become increasingly important with the successful control of macrofouling. b) Diatoms are the major constituents of ship microfouling films. c) A limited number of diatoms species/genera are prevalent foulers, and these diatoms exhibit a range of successful attachment mechanisms. d) The composition of a fouling film dictates its surface profile, and e) Rougher surface profiles are generally exhibited by thicker fouling films, which may lead to greater fuel consumption.

MICROBIAL FOULING OF METAL SURFACES

Introduction

Copper is an important biocide and extensively used in the marine environment as an antifouling agent, most commonly in paints as cuprous oxide. In alloy form, usually with nickel, it has the advantage over other metals because of its increased corrosion resistance whilst exhibiting antifouling properties, partly due to copper ions leaching from the surface (Moreton and Glover, 1980). For this reason copper alloys, particularly 90:10 cupro-nickel are used quite extensively in the marine environment as cladding on offshore structures such as gas platforms and for the construction of condenser tubes and fish net cages (Moreton and Glover, 1980). A programme of study was, therefore, initiated to determine the process of microfouling on copper alloys, in comparison with other metals such as titanium, and to examine the antifouling mechanisms.

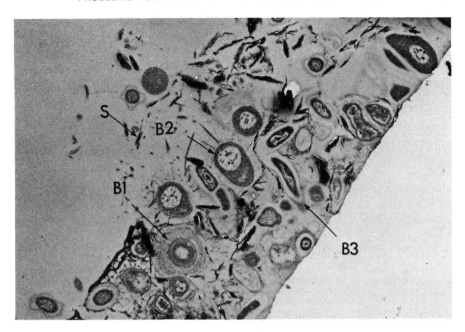

FIG 9 Transmission electron microscope micrographs showing slime film on titanium after 1 week exposure in Langstone Harbour with bacteria surrounded by a thick mucilagenous capsule (B1), bacteria producing mucilagenous strands (B2) and bacteria partially enclosed within a thin mucilagenous sheath (B3). Silt particles are found attached to the bacterial mucilage (S). × 13800.

Materials and Methods

The colonization of the metal surfaces exposed in Langstone harbour, south coast of England, was monitored using transmission electron microscopy (TEM). Four different surfaces were used: a) Commercially pure titanium b) Pure copper c) Aluminum brass alloy and, d) 90:10 cupro-nickel alloy. All metals were immersed as small discs in Langstone harbour from test rafts for different periods of time and fixed for TEM in 4% gluteraldehyde, 3% NaCl and 0.1 M sodium cacodylate buffer at pH 7.2. After post fixation in 2% osmium tetroxide in 0.1 M sodium cacodylate buffer (at pH 7.2) for two hours, the discs were dehydrated and embedded in Epon-araldyte epoxy resin mixture. The metal or alloy surface was stripped from the polymerized resin by immersion in liquid nitrogen and then in water at 40°C. Sections were cut through the corrosion products and the slime film using a diamond knife on an LKB 3 Ultramicrotome. After staining in uranyl acetate and lead citrate sections were observed in a JEOL 100S transmission electron microscope.

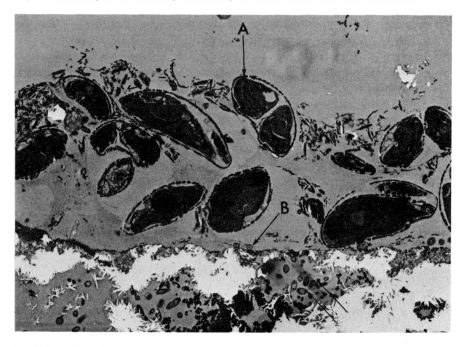

FIG 10 Slime film on aluminium brass after 15 weeks' exposure. This film is composed of bacteria (B) and the diatom *Amphora* (A). Bacteria (arrowed) are also found amongst the corrosion products (C). × 4600.

Results and Discussion

Titanium fouled relatively quickly when compared to the copper-based alloys. A bacterial film formed on the surface of titanium after one week (Figure 9). This film measured 4 μm thick and was composed of at least three different morphological types of bacteria. Some bacteria in this slime film were invested by a thick mucilagenous capsule (B1 Figure 9) whilst others produced strands of polymeric material or were partially ensheathed by mucilage (B2, B3 Figure 9). Silt particles were found buried or trapped by the mucilage produced by these various types.

In contrast to titanium, aluminium brass surfaces were first colonized by bacteria and diatoms after 15 weeks' exposure (Figure 10). Bacteria were found within the slime film and were also buried amongst the corrosion products produced by this alloy. The film which measured 12 μm thick, was dominated by the diatom *Amphora* (a common genus found on copper containing surfaces). Silt particles were also found attached to the mucilage produced by these cells.

Of the metals and alloys investigated copper was the most toxic in that it took up to 23 weeks for the copper discs to be colonized by the bacteria.

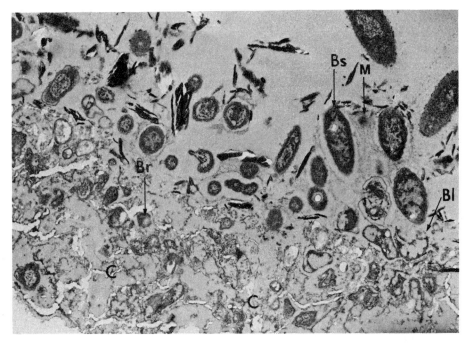

FIG 11   Bacteria on copper after 30 weeks' exposure. Showing healthy bacteria in slime film (Bs) producing mucilage (M), bacteria which are adjacent to the metal surface (B1), and remnants of bacteria (Br) in the corrosion products (Bc). × 18400.

These bacteria were observed within the corrosion products (Figure 11), however a true slime film was not detected above these corrosion products until 30 weeks' exposure. Bacteria at the surface of the slime film produced large amounts of mucilage and as viewed in the TEM their cell contents appeared to be normal, whilst those adjacent to the corrosion products were often lysed or abberant (B1 Figure 11). Remnants of bacteria could be identified within the corrosion products (Br Figure 11).

A layering arrangement was observed on 90:10 cupro-nickel after 19 weeks of exposure (Figure 12). Sheets of bacteria one cell thick were found sandwiched between parallel layers of corrosion products (Figure 12). After longer periods of exposure the surface of these corrosion layers were colonized by diatoms and brown filamentous algae to form slime films approximately 100 μm thick. When the samples were removed from the sea after stormy periods the surface appeared to be bright and devoid of any slime film. In the TEM there were few parallel layers present and it appeared that the antifouling properties can be attributed to the sloughing off of some of the parallel layers observed in Figure 12.

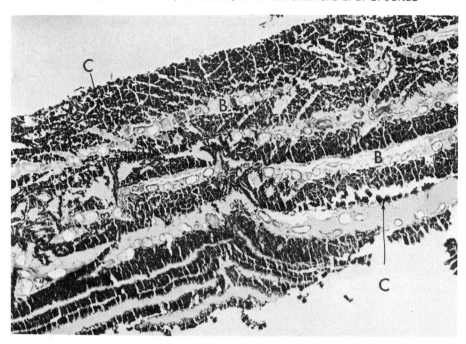

FIG 12   Parallel corrosion layers on 90/10 copper-nickel (C) with sheets of bacteria
sandwiched between the corrosion layers (B). × 6600.

Conclusions

In conclusion titanium fouled more readily and developed thicker slime
films than copper and copper alloys. This is explained by the fact that
titanium does not corrode in sea water while copper and copper alloys
produce toxic copper ions which prevent fouling, and in addition produce
corrosion products which slough off with a consequent loss of the microbial
film.

GENERAL CONCLUSIONS

In this paper a number of topics have been investigated and the results
illustrate the problems that arise when surfaces are exposed in the sea and
become colonized by micro-organisms. Not only do they colonize inert
materials, such as titanium, but also toxic surfaces, such as copper nickel
alloys and antifouling coatings. The net result is a loss of efficiency in
operating a range of marine facilities often with increased financial
expenditure.

ACKNOWLEDGEMENTS

We gratefully acknowledge financial support from the Ministry of Defence, Procurement Executive, the Research Organization of Ships Compositions Manufactures (R.O.S.C.M), and the International Copper Research Association Inc. We also wish to thank Mr E. Tiller for technical assistance during the heat transfer trials.

REFERENCES

Adamson, W. L., Liberator, G. L. and Taylor, D. W. (1984). 'Control of microfouling in ship piping and heat exchanger systems', in *Marine biodeterioration: An interdisciplinary study*. (eds. J. D. Costlow and R. C. Tipper), pp. 95–100, Naval Inst. Press, Annapolis.

Bacon, G. B. and Taylor, A. R. A. (1976). 'Succession and stratification in benthic diatoms colonising plastic collectors in Prince Edward Island estuary', *Botanica Mar.*, **29**, 231–240.

Baier, R. E. (1973). 'Influence of the initial surface condition of materials on bioadhesion', in *Proc. 3rd Int. Congr. Mar. Corr. and Fouling*, (eds. R. F. Acker, B. Floyd Brown and J. R. DePalma), pp. 633–639, Northwestern University Press, Evanston.

Baier, R. E. (1984). 'Initial events in microbial film formation', in *Marine biodeterioration: An interdisciplinary study*. (eds. J. D. Costlow and R. C. Tipper), pp. 57–62. Naval Inst. Press, Annapolis.

Barnes, H. (1970). 'A review of some factors affecting settlement and adhesion in the cyprid of some common barnacles', in *Adhesion in biological systems*. (ed. R. R. Manley), pp. 89–111, Academic Press, New York.

Bishop, J. H. (1969). 'The application of scanning electron microscopy to antifouling paint research', *Aust. OCCA Proc. and News*, **6**, 13–16.

Bishop, J. H., Marson, F. and Silva, S. R. (1972). 'Microfouling on antifouling coatings', *Aust. OCCA Proc. and News*, **9**, 4–5.

Bishop, J. H., Silva, S. R. and Silva, V. M. (1974). 'A study of microfouling on antifouling coatings using electron microscopy'. *J. Oil Col. Chem. Assoc.*, **57**, 30–35.

Blunn, G., and Jones, E. B. G. (1985). 'Antifouling properties of copper and copper alloys'. Copper Developmental Association. Paper 15, Edgebaston, April 1985.

Brown, I. and Jones, E. B. G. (1982). 'An investigation into microbial colonisation of the surface of titanium condenser tubes exposed to Thames river water', *Int. Biodet. Bull.*, **18**, 67–79.

Bryers, J. D. and Characklis, W. G. (1979). 'Measurement of primary biofilm formation', in *Condenser Biofouling Control Symp. E.P.R.I. CS-1450* (eds. J. F. Garey, R. M. Jorden, A. H. Aitken, D. T. Burton and R. H. Gray), Ann Arbor Sci. Publ. Inc.

Callow, M. E., Wood, L. V. and Evans, L. V. (1978). 'The biology of slime films; Part 4', *Shipp. Wld.*, 3939, 273–276.

Caron, D. A. and Sieburth, J. M. (1981). 'Distribution of the primary fouling sequence on fibre glass reinforced plastic submerged in the marine environment', *Appl. Env. Microbiol.* **41**, 268–273.

Characklis, W. G. (1973a). 'Attached microbial growths I', *Water Res.*, **7**, 1113–1128.

Characklis, W. G. (1973b). 'Attached microbial growths II', *Water Res.*, **7**, 1249–1259.

Characklis, W. G. (1979). 'Biofilm development and destruction in turbulent flow', in *Ozone: Science and Engineering*, **1**, 167–181.

Characklis, W. G. (1981). 'Fouling biofilm development; A process analysis', *Biotechnol. Bioeng.*, **23**, 1923–1960.

Christie, A. O., (1982). 'The economics of docking intervals, conventional antifoulings and advanced paint systems', *Shipcare, Marit. Manage.*, **14**, 25–31.

Clitheroe, S. B., (1983). 'From antifouling to antiroughness coatings', *Aust. OCCA Proc. and News*, **20**, 6–10.

Conn, A. F., Rice, M. S. and Hagal D. (1977). 'Ultra clean heat exchangers. A critical O.T.E.C requirement', in *Proc. 4th O.T.E.C. conf.*, pp. VII, 11–14.

Corpe, W. A. (1970a). 'Attachment of marine bacteria to solid surfaces', in *Adhesion in biological systems*, (ed. R. S. Manly), pp. 73–87, Academic Press, New York.

Corpe, W. A. (1970b). 'An acid polysaccharide produced by a primary film forming marine bacterium', *Dev. Ind. Microbial*, **11**, 402–412.

Corpe, W. A. (1973). 'Microfouling: the role of primary film forming marine bacteria', in *Proc 3rd Int. Congr. Mar. Corr. and Fouling*, (eds. R. F. Acker, B. Floyd Brown and J. R. DePalma), pp. 589–609, Northwestern University Press, Evanston.

Corpe, W. A. and Winters, H. (1972). 'Hydrolytic enzymes of some periphytic marine bacteria', *Can. J. Microbiol.*, **18**, 1483–1490.

Daniel, G. F. and Chamberlain, A. H. L. (1981). 'Copper immobilisation in diatoms', *Botanica mar.*, **24**, 229–243.

Daniel, G. F., Chamberlain, A. H. L. and Jones, E. B. G. (1980). 'Ultrastructural observations on the marine fouling diatom *Amphora*', *Helgolander Wiss. meeresunters.*, **34**, 123–149.

Dempsey, M. J. (1981). 'Colonisation of antifouling paints by marine bacteria', *Botanica mar.*, **24**, 185–191.

DePalma, V. A., Goupil, D. W. and Akers, C. K. (1979). 'Field demonstration of rapid microfouling in model heat exchangers', in *Proc. Ocean Therm. Energy Convers. Conf: Sixth Conf.*, Vol. 2, 12.14-1–12.14-9.

Evans, L. V. (1981). 'Marine algae and fouling: A review, with particular reference to ship fouling', *Botanica mar.*, **14**, 167–171.

Fischer, E. C., Castelli, A. J., Rodgers, S. D. and Bleile, H. R. (1984). 'Technology for the control of marine biofouling', in *Marine biodeterioration: An interdisciplinary study*, (eds. J. D. Costlow and R. C. Tipper), pp. 265–299, Naval Inst. Press, Annapolis.

Fletcher, R. L. (1980). *Catalogue of main marine fouling organisms. Vol. 6*, Algae, O.D.E.M.A., Brussels.

Fletcher, R. L., and Chamberlain, A. H. L., (1975). 'Marine fouling algae', in *Microbial Aspects of the Deterioration of Materials*, (eds. D. W. Lovelock and R. J. Gilbert), pp. 59–81, Academic Press, London and San Francisco.

Fowler, H. W. and McKay, A. J. (1979). 'The measurement of microbial adhesion', in *Microbial Adhesion to Surfaces* (eds. R. C. W. Berkeley, J. M. Lynch, J. Melling, P. R. Rutter and B. Vincent), pp. 143–161, Ellis Horwood Ltd., Chichester.

Garey, J. F., Jorden, R. M., Aitken, A. H., Burton, D. T. and Gray, R. H. (1980). *Condenser Biofouling control symposium proceedings*, Ann Arbor Science, Ann Arbor, MI.

Gerchakov, S. M., and Udey, L. R. (1984). 'Microfouling and Corrosion', in *Marine biodeterioration: An interdisciplinary study* (eds. J. D. Costlow and R. C. Tipper), pp. 82–87, Naval Inst. Press, Annapolis.

Gillmer, T. C. (1970). *Modern ship design*, Naval Inst. Press, Annapolis.

Hendy, N. I. (1974). 'A revised check-list of British marine diatoms', *J. mar. biol. Assoc. U.K.*, **54**, 277–300.

Huang, R., and Boney, A. D. (1983). 'Effects of diatom mucilage on the growth and morphology of marine algae', *J. exp. mar. Biol. Ecol.* **67**, 79–89.

Leitch, M. (1980). 'Subsea fouling; mussel-bound', *Petroleum Review*, Feb. 26–29.

Loeb, G. I. and Neihof, R. A. (1975). 'Marine conditioning films', in *Applied chemistry at interfaces*, Advances in Chemistry Series, No. 145, pp. 319–335, American chemical society, Washington.

Loeb, G. I. and Smith, N. (1981). 'Slime analysis of painted steel panels immersed in Biscayne Bay, Miami Beach, Florida', *N.R.L. Memo Report No. 4411*, Washington.

Marshall, K. C., Stout, R. and Mitchell, R. (1971a). 'Mechanism of the initial events in the sorption of marine bacteria to surfaces', *J. Gen. Microbiol.* **68**, 337–348.

Marshall, K. C., Stout, R. and Mitchell, R. (1971b). 'Selective sorption of bacteria from seawater', *Can. J. Microbiol.*, **17**, 1413–1416.

McDonnell, E. A., Mulder, J., Hodgekiss, T. and Boney, A. D. (1984). 'Some corrosion effects of marine micro-organisms', in *U.K. Corr., Proc. of Conf.* pp. 1–10, Publ. Inst. of Corr. Sci. and Tech.

Mitchell, R. and Kirchman, D. (1984). 'The microbial ecology of marine surfaces', in *Marine biodeterioration: An interdisciplinary study*, (eds. J. D. Costlow and R. C. Tipper), pp. 49–56, Naval. Inst. Press, Annapolis.

Millar, R. H. (1968). 'Ascidians as fouling organisms', in *Marine Borers, fungi and fouling organisms of wood, Proc. O.E.C.D. Workshop*, (eds. E. B. G. Jones and S. K. Eltringham), pp. 185–191, O.E.C.D. Paris.

Moreton, R. M. and Glover, T. J. (1980). 'New marine industry applications for corrosion and biofouling resistant, copper-nickel alloys', in *Proc. 5th Int. Congr. Mar. Corr. and Fouling*, pp. 267–278, Editorial Garsi, London and Madrid.

Moss, B. and Woodhead, P. (1970). 'The breakdown of paint surfaces by *Enteromorpha* sp.', *New Phytol.*, **69**, 1025–1027.

Nelson Smith, N. (1967). *Catalogue of main marine fouling organisms, Vol. 3, Serpulids*, O.E.C.D. Paris.

Nimmons, M. J. (1979). *Heat transfer effects in turbulent flow due to biofilm development*, Masters Degree, Rice University, Houston, Texas.

Norrman, G., Characklis, W. G. and Bryers, J. D. (1977). *Control of microbial fouling in circular tubes with chlorine*, Dev. Ind. Microbiol., **18**, 581–590.

O'Neill, T. B. and Wilcox, G. L. (1971). 'The formation of a primary film on materials submerged in the sea at Port Hueneme, California', *Pac. Sci.*, **25**, 1–12.

Preiser, H. S., Ticker, A. and Bohlander, G. S. (1984). 'Coating selection for optimum ship performance', in *Marine biodeterioration: An interdisciplinary study* (eds. J. D. Costlow and R. C. Tipper), pp. 223–229, Naval Inst. Press, Annapolis.

Pyne, S., Fletcher, R. L. and Jones, E. B. G. (1986). 'Diatom communities on non-toxic substrata and two conventional antifouling surfaces immersed in Langstone Harbour, South coast of England', in *Algal Biofouling*, (eds. L. V. Evans and K. D. Hoagland), Elsevier, Amsterdam, (in press).

Robinson, M. G., Hall, B. D. and Voltolina, D. (1985). 'Slime films on antifouling paints. Short-term indicators of long-term effectiveness', *J. Coatings Technol.* **57**, 35–41.

Ryland, J. S. (1965). *Catalogue of main marine fouling organisms (found on ships coming into European waters), Vol 2, Polyzoa, O.E.C.D.*, Paris.

Somerscales, E. F. C. and Knudsen, J. G. (1981). 'Fouling of heat transfer equipment' Hemisphere, New York.

Southward, A. J. and Crisp, D. J. (1963). *Catalogue of main marine fouling organisms, Vol 1, Barnacles*, O.E.C.D. Paris.

Terry, L. A. and Edyvean, R. G. J. (1981). 'Microalgae and corrosion', *Botanica mar.*, **24**, 177–183.

Thomas, R. W. S. P. and Allsop, D., (1983). 'The effects of certain periphytic marine bacteria upon the settlement and growth of *Enteromorpha*, a fouling algae', *Biodeterioration*, **5**, 348–357.

Young, L. Y. and Mitchell, R. (1973). 'The role of microorganisms in marine fouling', *Int. Biodetn. Bull.*, **9**, 105–109.

BIOCIDES FOR THE OIL INDUSTRY

K. D. Brunt

The Boots Company PLC

INTRODUCTION

This review about biocides for the oil industry is intended to encourage some real consideration of the properties of these chemicals and the ways in which they can best be used. Biocide producers and the service companies that formulate biocidal chemicals into products for the offshore oil industry feel that the true end-users do not appreciate fully the wide range of their applications. Even the large oil operators can be accused of not always attaching adequate performance specifications to their orders or tenders for supply of biocides.

The problem may well begin with the field operators who may fail to acknowledge the need to take preventative measures against the problems that can be caused by micro-organisms. The many Canadian gas wells which have become sour over the past few years provide evidence for such a lack of foresight and the use of biocides is increasingly being required. Recession current in the Canadian oil industry means that oil operators can no longer afford to dump batches of spoiled drilling mud. It is also desirable to avoid undue capital outlay on H_2S scrubbing equipment and other environmental clean-up measures now being demanded by the federal authorities.

It has now been accepted in Canada that the only way to ensure a long and efficient production life in oil or gas wells is to take appropriate antimicrobial action from the outset of drilling operations throughout the life of the system.

In the North Sea, although great efforts have been made by the majority of operators to control down-hole contamination, one or two small fields have already gone sour. It is relevant, therefore, to examine those biocides available for use in offshore oil operations.

BIOCIDES

The data on biocides given in this review represent some of the major antimicrobial compounds used in the legion of biocidal formulations available

201

Table 1   The main chemical groups of biocides used in
offshore oil operations

Biocide type	Manufacturers*
Aldehydes	Shell
	Union Carbide
Biguanides	ICI
Bromine/nitrogen cpds	Angus
	Boots
	Buckman
	Dow
Isothiazolones	ICI
	Rohm & Haas
Quaternary ammonium cpds	ABM
	Akzo
	Bayer
	Ethyl
	Lonza
	Onyx
Thiocyanates	Buckman
	Merck
	Stauffer
Other	BDH
	Givaudan
	Dow

*This does not constitute an exhaustive list.

to the oil industry from the various service companies. It is intended to give
the oil producer a better concept of the basic attributes of the different
active compounds available which, in turn, will help them to make more
specific and informed requests for oilfield biocide formulations.

An exhaustive list of biocides used in offshore operations would be almost
impossible to compile; novel products or new formulations of existing
biocides are continually being presented to the industry for the many
different applications that may require treatment. By way of compromise
a list of the different groups of chemicals with biocidal activity used in the
offshore oil production industry is given in Table 1 together with examples
of manufacturers or suppliers of biocides based on these antimicrobial
compounds. Addresses of many of the biocide companies are listed in Table
2. The reader is asked to approach these and other companies directly for
details of their products. (Mention of any chemical does not imply
endorsement for its use by the author or by the publishers).

Many, if not most, biocides need careful handling, especially when in
a concentrated form. Even at in-use levels, some biocides may be irritant
or cause skin sensitisation effects if not used as recommended. Disposal
of biocide-treated waters or other fluids must take into consideration

Table 2   Reference list of biocide manufacturers/suppliers

ABM Chemicals Ltd., *Woodley, Stockport, Cheshire. SK6 1PQ*
AKZO Chemie, *1–5 Queens Road, Hersham, Walton-on-Thames, Surrey. KT12 5NL*
Angus Chemical Co., *19 Moorgate Street, Rotherham. S60 2DA*
Bayer UK Ltd., *Bayer House, Strawberry Hill, Newbury, Berks. RG13 1JA*
BDH Chemicals, *Broom Road, Poole, Dorset. BH12 4NN*
The Boots Company PLC, *Thane Road, Nottingham. NG2 3AA*
Buckman Labs., *Wondelgemkaai 159, B-9000, Ghent, Belgium.*
Dow Chemical Co. (UK), *Meadowbank, Bath Road, Hounslow, Middlesex. TW5 9QY*
Ethyl Corp., *330 South Fourth Street, Richmond, VA 23219, USA.*
Givaudan, *50–56 Rue Paul-Cazeneuve, BP 8344 Lyon 08, 69356 Lyon Cedex 2, France.*
Imperial Chemical Industries Ltd., *Organics Division, P.O. Box 42, Blackley, Manchester. M9 3DA*
Lonza (UK) Ltd., *Imperial House, Lypiatt Road, Cheltenham, Glos. GL50 2QJ*
Merck Holding Ltd., *Broom Road, Poole, Dorset.*
Onyx Chemical Co., *Newark, New Jersey, USA.*
Rohm & Haas (UK) Ltd., *Lenning House, 2 Mason's Ave., Croydon. CR9 3NB*
Shell International, *Shell Centre, London. SE1 7NA*
Stauffer Chemical Co., *Westport, CT 06881, USA.*
Union Carbide, *Union Carbide House, High Street, Rickmansworth, Herts. WD3 1RB*

---

whether the product is toxic to the environment. Safety and ecotoxicological data are, therefore, as vital to the choice of biocide as efficacy and compatibility with the systems treated.

ACTIVE INGREDIENTS OF BIOCIDES

Aldehydes

Acrolein is not widely used in the North Sea but has been employed as a biocide in the Middle East. However, although effective antimicrobially (minimum inhibitory concentrations less than 1 ppm) it is highly unstable and unsuitable for handling as a biocide being irritant and toxic by skin absorption.

Glutaraldehyde is much more widely used. It is reasonably safe to handle and very effective at 5–80 ppm against a broad range of bacteria. It is used in water injection systems, workover and completion fluids, fracturing fluids and drilling fluids at 10–50 ppm (active ingredient). It is readily water soluble and stable in aqueous solution above pH 3.7. A drawback is a tendency to be inactivated by organic matter and so glutaraldehyde is not advisable for use in heavily slimed systems.

Biguanides

The cationic polymeric biguanides are not widely used in oil field operations but are safe compounds to handle with useful activity at 500–2000 ppm

Table 3   Evaluation of sulphate-reducing bacteria in produced
water before and after treatment with 2-bromo-2-nitropropane-
1,3-diol (97% active ingredient)

| Well no. | Sulphate-reducing bacteria levels | |
	Before treatment	7 days after treatment
1	+ (High levels)	−
2	+ (High levels)	+ (Reduced levels)
3	+ (High levels)	−
4	+	−
5	−	−
6	+ (Low levels)	−
7	+	+ (Reduced levels)
8	+	+ (Reduced levels)
9	+	No water
10	−	−
11	+ (Low levels)	−
12	+	+ (Reduced levels)

for some water treatment applications. They are readily soluble in water
and aliphatic alcohols but generally insoluble in hydrocarbons and aromatic
solvents. Biguanides are stable over a wide pH range.

Oxazolidines

These heterocyclic compounds are effective broad spectrum biocides
(Minimum Inhibitory Concentration − MIC, 60–250 ppm) and have been
recommended for intermittent or continuous use in injection water systems
at 5–150 ppm.

Bromine Nitrogen Compounds

A number of different biocidal chemicals are included in this miscellaneous
group. In North Sea oil operations, the compound 2-bromo-2-nitropropane-
1,3-diol has been used in water treatment, in drilling muds and for downhole slug dose treatments. It is soluble in water and polar solvents but
insoluble in nonpolar solvents. Stability is adversely affected with increase
in pH (above pH 6.5) and at higher temperatures. It is active against a wide
range of organisms (MIC 1–50 ppm). Examples of its use in oil field
treatments are given in Tables 3 and 4.

Another compound in this group 2-hydroxyethyl-2,3-dibromopropionate,
is effective over a wide pH range and has been used to treat injection water
and produced water at levels of 8–30 ppm.

Table 4 Effect of treatment of drilling fluids during gas well trials with 97% (active ingredient) 2-bromo-2-nitropropane-1,3-diol

Systems tested	Parameters measured	Microbial counts (aerobic bacteria)							
		Treated wells					Untreated wells		
		1	2	3	4	5	6	7	8
Fluid before drilling		4×10^1	4×10^1	4×10^1	4×10^1	4×10^1	3×10^3	1.4×10^6	2×10^5
Fluid after drilling	Depth 350 m	1×10^2	0	0	3×10^2	9.4×10^2	1×10^2	2.4×10^4	
	550 m	No sample	1.6×10^2	No sample	0	$*1.7 \times 10^3$			

*This result is suspect due to questionable sampling technique.

The biocide 2,2-dibromo-3-nitrilopropionamide is active against a wide spectrum of bacteria although it loses biological activity above pH 9. It decomposes on heating.

Isothiazolones

The isothiazolones have powerful antibacterial activity but they require careful formulation to give stability and water solubility. Like many biocides, they need fairly careful handling. They are used in the control of biofouling and in preservation. The isothiazolones are, however, often incompatible with corrosion inhibitors and are, in common with many other biocides, inactivated by systems containing H_2S.

BIT (1,2-benzisothiazolin-3-one) has been used in drilling muds and in water treatment in oil operations. It has a broad spectrum of activity (MIC 1-10 ppm) and is stable at in-use levels at pH 7.5 and above, decomposing at lower pH. BIT is miscible with water in all proportions.

A mixture of 5-chloro-2-methyl-4-isothiazolin-3-one and 2-methyl-4-isothiazolin-3-one is highly active (as low as 1 ppm) and has been used in drilling muds and water systems. It is miscible with water and most stable at alkaline pH.

Quaternary Ammonium Compounds

As biocides, the quaternary ammonium compounds (QACs) have been in use for many years but have only limited application in the offshore oil business. They are stable over a wide pH range but their surfactant characteristics can lead to foam problems in the treated systems. They are generally safe to handle but may be toxic to marine life.

Thiocyanates

The thiocyanates, traditionally used in many aspects of water treatment, are used as injection water biocides. Stability and activity decrease as pH increases.

Miscellaneous compounds

Dichlorophen is a highly stable, safe compound which has been used in drilling muds, packer and completion fluids and in diving suits. It is active against a wide range of organisms. Another safe and stable biocide recently developed is based on 2-hydroxyethoxyl)-methane. The very stability of such biocides may bring problems when treated systems are discharged into the environment. The hexamine derivative, 1-(3-chloroallyl)-3,5,7-triaza-1-azoniaadamantane chloride, has a good spectrum of activity and has been used in the treatment of water systems.

CONCLUSIONS

It is incumbent on biocide users, whether they are oil producers or service companies, to define clearly the purpose for which the biocide they are seeking will be used. The following points should first be identified:

What are the target groups of micro-organisms?
Where are these organisms located in the system?
What are the physical and chemical conditions of the system to be treated?
What are the available biocide dosing points?
What are the financial consequences of the contamination problem?

A biocide is required to have:

The correct spectrum of activity for the micro-organisms in question.
The right properties to enable it to be effective within the operating environment.
The correct formulation for the dosing points available.
Adequate safety for protection of the workforce.
An ecotoxicological pattern that is consistent with the local environment of the system.
A comprehensive data package.
A manufacturer who will provide a comprehensive support package to the use of his product.
A price which will show a saving, if used, over estimated costs due to contamination or biodeterioration.

Success in marketing of biocides combines a highly active product with professional, technically oriented market support operations. However, the biocide manufacturer or supplier needs to be given adequate and accurate information about the system his product is being called upon to treat if the most suitable biocide, formulation or dosing regime can be selected.

ACKNOWLEDGEMENTS

The author would like to thank those manufacturers who kindly sent their latest literature, accompanied in some cases by most helpful comments.

The opinions expressed in this paper are personal to the author and should not be taken as expressing the official opinion of The Boots Co. PLC or of the Institute of Petroleum.

BIOCIDE APPLICATION AND MONITORING IN A WATERFLOOD SYSTEM

S. Maxwell, K. M. McLean and J. Kearns

*Corrosion Specialists (North Sea) Limited, Aberdeen,
Hamilton Brothers Oil and Gas Ltd, Aberdeen*

INTRODUCTION

Control of bacteria, in particular sulphate-reducing bacteria in waterflood systems is generally by means of regular slug dosing with non-oxidizing organic biocides. One of the most commonly used treatments is to dose weekly for 6 hours with 200 ppm of a 25% solution of glutaraldehyde or modifications thereof. Such treatments were designed based on laboratory evaluations of the biocide against planktonic bacteria. In recent years, however, the importance of biofilms in microbial corrosion, and the resistance of sessile bacteria to biocides, has been recognized (Rosser *et al.*, 1984; Hamilton & Maxwell, 1986; Maxwell, 1986). In the light of this evidence it is now realized that the recommended treatments may have little or no effect on bacterial biofilms and that much higher concentrations of biocide may be necessary. It has been suggested that the concentration rather than the residence time of the biocide treatment is the critical factor. Thus it may be possible to maintain a cost effective treatment by increasing the concentration of biocide but reducing the time over which the biocide is injected. Detailed monitoring of both planktonic and sessile bacteria is essential to ensure that biocide treatments are optimized.

This paper presents results of a detailed monitoring programme performed on a water injection system in the Duncan Field (Hamilton Brothers Oil and Gas Limited). The metal surfaces of the flowlines have been maintained free from bacterial fouling by the implementation of regular high concentration short residence time shock doses of glutaraldehyde.

Duncan Field Water Injection System

The Duncan Field is a small oilfield in the northern North Sea which is produced entirely by subsea wells to the floating production facility, the Deepsea Pioneer. Early in the development stage the need for water

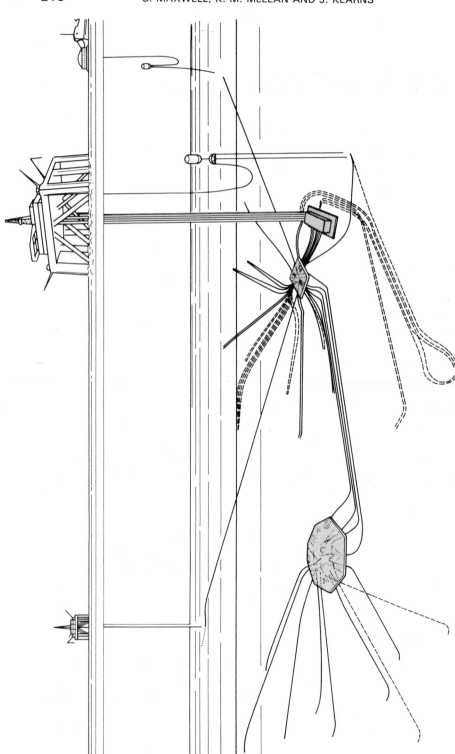

FIG 1   Argyll and Duncan Fields: Production layout.

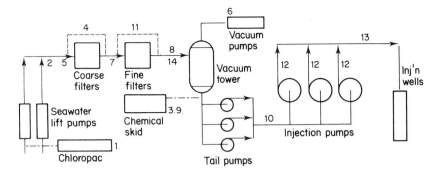

FIG 2   Schematic diagram of Deepsea Pioneer water injection system showing monitoring and chemical injection points.

1. Chloropac
2. Pressure, temperature, bacterial counts
3. Scale inhibitor injection
4. Coarse filter pressure drop
5. Coulter, Millipore, turbidity, chlorine
6. Pressure
7. Polyelectrolyte (not used)
8. Chlorine
9. Oxygen scavenger injection
10. Oxygen, temperature pressure
11. Fine filters pressure drop
12. Coulter, Millipore, turbidity, oxygen, residual sulphite, iron. Bacterial counts (SRB, GAB) before and after biocide addition
13. Injection pressure, flowrate
14. Biocide injection

injection to supplement aquifer pressure was identified and the opportunity taken to maintain cleanliness of the system from the outset of production.

The Duncan System is a relatively small injection project consisting of two subsea wells. The two wells are 7.4 and 5.6 kilometres distant respectively from the floating production facility (Figure 1). Corrosion control of the subsea injection lines, manifolds and downhole tubular structures is of the utmost importance due to the prohibitive cost of replacing failed equipment. The potential for bacterial growth and subsequent biologically mediated corrosion was identified at an early stage. Within the corrosion control programme a biological monitoring programme was initiated which would allow the most accurate monitoring of the topside flowlines, thereby allowing an accurate assessment of the degree of bacterial contamination further downstream.

A schematic diagram of the system, illustrating the various monitoring and chemical injection locations on the topside equipment, is shown in Figure 2.

BACTERIAL CONTROL

Bacterial control of the Duncan Field injection system is maintained by a two stage biocide treatment.

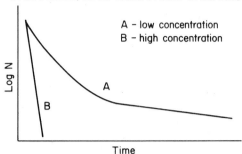

FIG 3   Effect of concentration on biocide killing.

Chlorination

Upstream of the deaerator, continuous chlorination is carried out. Chlorine is produced by hypochlorinators and injected into the injection water at the seawater lift pump inlet caissons. Residual chlorine levels upstream of the tower are maintained between 0.5–0.8 ppm continually. Chlorine levels are monitored twice a day and any shift above or below these levels is immediately rectified. The chlorine treatment maintains biological control within the coarse and fine filters and to some extent in the deaerator tower. Within the deaerator, however, the residual chlorine is inactivated by the addition of bisulphite oxygen scavenger and thus, downstream of the deaerator, a secondary biocide treatment is required.

Organic Biocide

Downstream of the deaerator slug dosing with an organic biocide is performed to maintain biological control. The choice of biocide was made based on published data and previous experience. Glutaraldehyde in 25% solution has been shown to be a very effective biocide against a wide range of marine bacteria. Numerous tests have been performed with glutaraldehyde against planktonic bacteria isolated from offshore injection systems. The results of these tests have formed the basis of the common slug treatment; 200 ppm 25% glutaraldehyde for 6 hours once per week, and slight variations thereof.

More recent laboratory and field tests of biocides against sessile bacteria have indicated that, in order to be effective against bacteria attached to surfaces, a glutaraldehyde concentration of 1000–2000 ppm (250–500 ppm active ingredient) may be necessary. Concentration has been indicated as the critical factor (Figure 3). Thus much reduced residence times can be employed, allowing the treatment to remain cost effective.

Initial budgeting allowed for weekly slug doses of 200 ppm 25% glutaraldehyde for 6 hours. In order to stay within this budget a recommendation

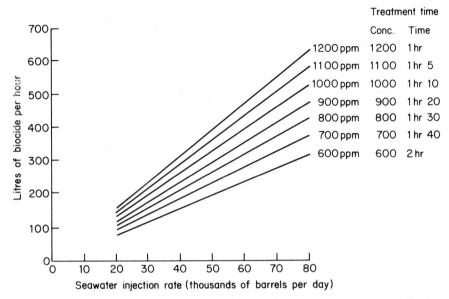

FIG 4   Maximum biocide concentration achievable at a range of water injection rates.

to dose biocide at 1200 ppm for 1 hour per week was made. Initial injection rates were 60,000 barrels per day and, with the pumps available in the chemical skid, a dose of only 900 ppm could be achieved. The residence time was, therefore, increased to 1 hour 20 minutes. A graph was prepared (Figure 4) showing the maximum biocide concentration achievable at a range of water injection rates. As the chemical pumps were always set at maximum pumping rate the residence time of any treatment was related directly to the water injection rate.

Many biocides include a surfactant to enable penetration of the biocide into the biofilm to achieve a more efficient kill. In this case, however, as the system was new, it was not felt necessary to use a surfactant biocide as no significant biofilm formation would have occurred prior to start-up.

START-UP PROCEDURE

Prior to injection start-up, specific chemical treatments were dosed through the long subsea flowlines in order to ensure the lines were free from biological and solid debris before water injection commenced.

Each line was treated with a slug of gel cleaner to remove solid debris prior to a 2400 ppm slug of 25% glutaraldehyde. Normally the system would be shut in with biocide for a period of 24 hours prior to start-up. In this case, to avoid a costly shut down of the plant, a high concentration

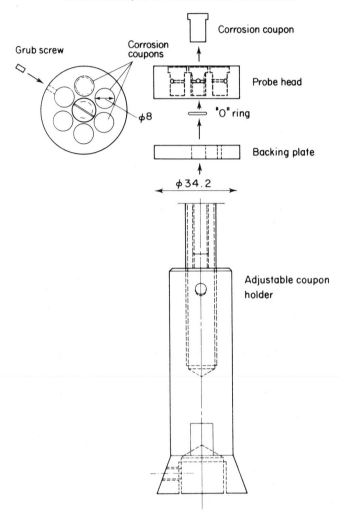

FIG 5   The bioprobe device.

slug dose of biocide was used. In order to achieve the desired effect it was calculated that this concentration need only be maintained in the line for a short period and a 30 minute treatment was administered.

BACTERIAL MONITORING

Two distinct monitoring programmes are operated to ensure that bacterial growth in the water injection system is being controlled. Routine weekly monitoring is performed by enumerating planktonic bacteria in the bulk

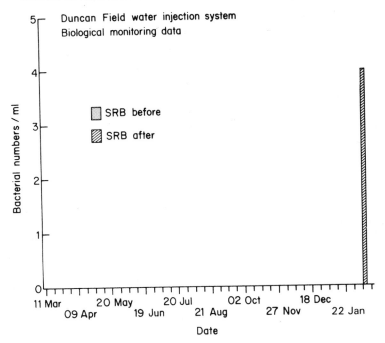

FIG 6 Number of sulphate-reducing bacteria (SRB) before and after biocide
treatment.

phase, whilst more detailed sessile monitoring is performed using a bioprobe
device (Figure 5) and surface scrapings every three months. Both monitoring
programmes are continually being reviewed and updated. The most recent
addition to the monitoring programme is the addition of a side stream
apparatus.

Planktonic Monitoring

It is widely realized that planktonic monitoring alone indicates very little
about the system being operated. Frequent sampling, however, when
coupled with detailed cross reference to other chemical and corrosion
monitoring data, allows trends to be detected in bacterial levels which will
aid the overall monitoring programme.

Weekly monitoring is performed, both before and after the biocide
treatment, at a sample point downstream of the injection pumps. At present
only sulphate-reducing bacteria (SRB) are monitored weekly, using the serial
dilution Most Probable Number (MPN) method with triplicate dilutions. Due
to the total absence of SRB in this monitoring programme, monitoring of
heterotrophic bacteria has now been introduced on a weekly basis.

Once per month a microbiologist visits the field to perform more detailed monitoring including the use of filtration techniques for recovery of SRB and sampling upstream and downstream of the deaerator for both SRB and hetero-trophic bacteria (commonly referred to as General Aerobic Bacteria (GAB)).

At this time the chemical injection pumps are recalibrated. A careful record of the chemical usage is kept to ensure that the system is being neither undertreated, in which case the system is at risk, nor overtreated causing wastage of expensive chemical.

Sessile Monitoring

Sessile bacterial monitoring is the only accurate means of ensuring that bacterial biofilm formation is being controlled. After five months of operation a bioprobe device (Figure 5) was installed in the water injection header. At this time surface scrapings of the pipewall were collected and tested for the presence of bacteria. No SRB and only very low levels of GAB were detected in these scrapings. After three months' exposure the bioprobe was removed from the system and the coupons examined for sessile bacteria by viable counts using serial dilution and direct epifluorescence counting. Due to the results obtained from previous monitoring it was not felt necessary at this time to perform more detailed analysis on the coupons (e.g. radiorespirometry) and this was confirmed by our results.

RESULTS

Planktonic Monitoring

The results of the routine planktonic testing are presented in Figures 6 and 7 and the results of the chlorine monitoring in Figure 8. The level of SRB in 1 ml samples has remained below detection. SRB can regularly be detected, however, by filtering one litre of injection water. This level remains constant both before and after the biocide treatment and it is, therefore, felt that this is a background level which cannot be improved upon. Only on one occasion have SRB been detected in 1 litre samples upstream of the deaerator. The level of chlorination at this time had dropped below the minimum acceptable limit of 0.5 ppm. During start-up the injection of glutaraldehyde was downstream of the deaerator and, therefore, some bacterial growth may have occurred in the deaerator at this time. Furthermore, when levels of chlorination dropped an immediate increase in levels of GAB was seen. Chlorination is now at acceptable levels (0.5–0.8 ppm) and, with biocide injection being upstream of the deaerator, a gradual decrease in GAB numbers is being seen. The levels of GAB detected at the injection and booster pumps are felt to be the result of contamination of the deaerator.

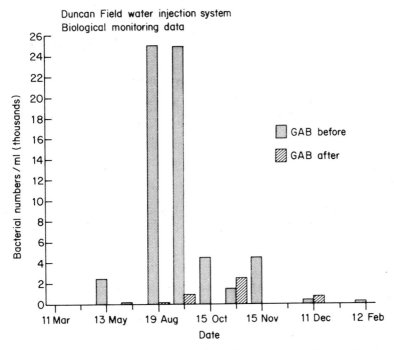

FIG 7   Numbers of general aerobic bacteria before and after biocide treatment.

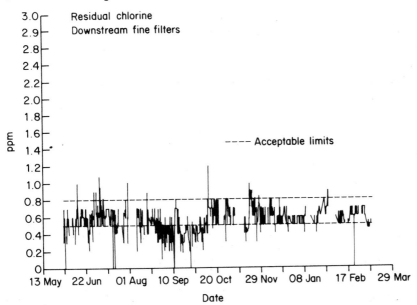

FIG 8   Residual chlorine levels downstream of fine filters. (Sample point 8 in Figure 2).

Table 1   Results of sessile monitoring from the Duncan Field water injection system

| Date | Sample | Viable counts | | Epifluorescence |
		SRB cm^{-2}	GAB cm^{-2}	TOTAL COUNT cm^{-2}
19/08/85	Surface scraping	0	2	ND[1]
14/01/86	Corrosion coupon	0	7.5×10^1	ND
14/01/86	Bioprobe stud	0	3.8×10^2	9.0×10^3

[1]ND test not done.

Sessile Monitoring

The results of the bioprobe and corrosion coupon monitoring are presented in Table 1. It can be seen that no significant biofouling has occurred.

CONCLUSION

The extended length of the subsea flowlines in the Duncan Field injection system means that the monitoring data gathered on the topside facilities must be extrapolated throughout the system. It is essential, therefore, that the results from the topside monitoring be the most accurate possible. Furthermore, these results must indicate complete control of the system topside if the chemical treatment is to be justified.

Frequent monitoring has highlighted the importance of the primary chlorination in overall bacterial control.

It is appreciated that once biofilms have become established it is extremely difficult to regain biological control of the system. Therefore, a high workload in maintaining the system free of biofouling is totally justified.

ACKNOWLEDGEMENT

The authors wish to thank Hamilton Brothers Oil and Gas Limited for the opportunity to present this paper.

REFERENCES

Hamilton, W. A. and Maxwell, S. (1986). 'Biological corrosion activities of sulphate-reducing bacteria within natural biofilms', *Proceedings International Conference Biologically Induced Corrosion*, 1985 (in press).

Maxwell, S. (1986). 'The assessment of sulphide corrosion problems by biological monitoring', *SPE J. Prod. Engng*, (in press).

Rosser, H. R., Purdy, R. L. and Roberson, G. R. (1984). 'Development of biocide application parameters for sessile bacteria control in a large seawater water flood system'. *Proceedings U.K. Corrosion Conference, November 1984*, pp. 139–145.

FUELS

E. C. Hill

*E. C. Hill and Associates, Unit M22,
Cardiff Workshops, East Moors, Cardiff*

INTRODUCTION

It has been known for ninety years that micro-organisms can feed on hydrocarbons but it is only in the last thirty years that the significance of this has been appreciated in terms of spoilage of petroleum products and accelerated corrosion. A comprehensive review was presented by Hill (1984) and a practical guide published for the marine industry by the General Council of British Shipping (Hill, 1983). Water is essential for microbial growth and although the potential for microbial growth exists in most water contaminated petroleum products the real risk fluctuates according to changes in the product formulation, the conditions of use and the amount of water present.

MICROBIOLOGY OF FUELS

In 1939 Thaysen attributed an explosion in a kerosene tank to gas generated by bacteria and spasmodic publications recorded microbiological incidents for the next two decades. The advent of the jet engine and the widespread use of kerosene fuels for it, precipitated a multitude of problems and proliferation of technical papers. The carbon chain length present (C_{10}–C_{18}) is much more readily assimilated by microbes than those in gasoline (C_5–C_9) and the lead anti-knock additives in gasoline are to some extent anti-microbial. The severe fouling and corrosion experienced in jet fuel has been extensively investigated and although anti-microbial fuel additives came into use the problems were largely resolved by good housekeeping. Our own investigations showed that microbial growth also occurred in tractor vapourizing oil, diesel, paraffin, gas oil etc but the consequences were not as dramatic as in aircraft. Hence minor fouling and corrosion were tolerated and the attention given in the aircraft industry to water removal and monitoring for microbes was not considered justified by other fuel users. The last few years have, however, seen a dramatic increase in severe microbial problems, particularly in the marine industry.

219

For this there must be a logical reason and we believe that this is firstly due to the recent changes in the chemistry of distillate fuels due to different procedures of refining and blending (Unzelman, 1984) and secondly to the widespread use of fuel additives, some of which stimulate microbial growth. The overall result is that many fuels are nowadays excellent food for micro-organisms.

For the first time in our laboratory we are detecting very high levels of microbial growth in heavier black fuel oils as well as the lighter distillate fuels.

Whereas at one time the fungus *Cladosporium resinae* was the *bete noir* of fuel users it has now been replaced by a very broad spectrum of bacteria, yeasts and fungi. Distillate fuels contain ample oxygen to sustain aerobic microbial growth but when tanks are stagnant this oxygen is depleted and anaerobic sulphate-reducing bacteria (SRB) such as *Desulfovibrio* can flourish. SRB never exist alone but always in consortia of many other organisms.

These consortia, which become established in the water phase are largely sustained nutritionally from the fuel phase across the interface, with some nutrients derived from the water according to its source. Fuel, even with its additives, is hardly a balanced diet but as it is continuously used and replenished (whilst the water and microbes remain *in situ*) an adequate nutrient supply is available.

In vehicle, ship and shore installations the problem is different from the better known jet aircraft problem in several respects. In the latter the wing temperature may fall to about $-30°C$ in flight and hence growth only occurs spasmodically when the aircraft is grounded and also the contaminating water is small amounts of pure condensate water rather than large volumes of sea or ground water. These factors influence the types of organisms and their rates of growth. In general a wider spectrum of organisms and a faster growth rate can be expected in non-aircraft fuel systems.

CONSEQUENCES OF MICROBIAL GROWTH

Small amounts of free water, however heavily infected, are unlikely to cause significant fouling of the fuel or a change in the fuel properties but could be associated with intense corrosion. However, the more water, the more biomass and the more problems.

We can summarize the consequences of microbial growth as follows:

(a) The physical presence of the organisms causes slimes and blocks pipes, valves and filters (Arnold, 1983).

(b) The surface activity of the organisms, particularly bacteria, promotes stable water in oil emulsions and causes coalescer malfunction. Fungi often continue to grow on coalescer cloth sleeving.

(c) Organisms foul fuel probes and cause incorrect volume measurement.
(d) Microbes accelerate corrosion by:
1. Producing corrosive products, particularly organic acids and sulphide. The latter dissolves in fuel which becomes corrosive and may cause it to fail copper or silver coupon tests.
2. Creating an oxygen gradient and hence an electro-chemical corrosion cell, resulting in anodic pitting. Creating other gradient corrosion cells.
3. Removing hydrogen from steel surfaces and thus depolarizing them.
4. Degrading corrosion inhibitors.
5. Degrading protective coatings such as paint, polyurethane and rubber.
6. Preventing the formation of stable oxide films on metals such as stainless steel and aluminium.

All of these consequences tend to be tolerated at minor levels of infection and are only investigated when failure or malfunction occurs. In a ship, vehicle or static fuel installation the most obvious signs of a high infection level are filter plugging, coalescer malfunction, injector fouling and corrosion. In ships, as the organisms are stimulated by warm temperatures, growth will be most rapid in the water-bottoms of service/settling tanks and less rapid in double-bottom tanks. Agitation and surfactancy distribute the microbes from the water bottom into the fuel phase where they are associated with minute water droplets.

Although theoretical growth rates predict rapid escalation of a problem, in practice there are limiting factors such as oxygen or phosphorus availability (organisms particularly require carbon, nitrogen and phosphorus as nutrients and a wide range of other elements in minor amounts). Thus we can expect that the time scale for a serious problem to develop is weeks rather than hours.

Because of the short contact time and the large bulk of fuel there are unlikely to be any detectable changes in its chemical composition.

DETECTION OF MICRO-ORGANISMS

The ideal sample for examination is a fuel water-bottom sample. This can be examined for living organisms by a competent laboratory both quantitatively and qualitatively using conventional cultural techniques. In general we can always expect some microbial contamination and hence even several thousand bacterial per ml is relatively insignificant. However, hundreds of thousands and more will suggest high levels of surfactancy and slime formation. Fungi tend to be less surfactant but as they grow in mats and copiously produce organic acids they can trap water, sediment and other organisms, block pipes, valves etc, and induce corrosion in metal in contact with the fungal mats. Numbers alone are not a good indication of fungal contamination as one unit may be several mm across or it may be simply a spore of only a few μm.

Thus microscopic examination of the sample is essential if any fungi are detected to determine the extent of growth. Yeasts are also copious acid producers and they grow prolifically in gas oil. As each yeast cell is an order of magnitude bigger than a bacterium (usual size $1-3\,\mu$m) smaller numbers of yeasts are more significant than the same numbers of bacteria. All of these organisms can selectively degrade various hydrocarbon components of the fuel and the fuel additives.

The presence of any sulphate-reducing bacteria indicates a long standing infection as they feed on the organic acid by-products of other organisms and need the other organisms to use up oxygen to which they are sensitive. Their presence suggests a high corrosion hazard. However, any SRB detected may have been derived from the fuel source and are not necessarily 'home grown'.

Water-bottom samples may only be readily available from settling tanks but water discharged from coalescers can also be a good sample.

If a water sample is not available a fuel sample can be taken and tested for living organisms by membrane filtration. An advised method for this has recently been demonstrated (Shennan, 1985) based on work co-ordinated by the Institute of Petroleum, London. However, the results from a membrane filtration test on fuel require careful interpretation. In a well settled tank there may be very few organisms in the fuel phase as biological material has a density a little greater than unity and may largely settle out over a few days. Obviously a fuel sample taken after a centrifuge, filter or coalescer will also be misleading as all these devices partially remove micro-organisms. In an agitated tank the ability of microbes to survive in fuel after passing into suspension there varies enormously. Bacteria may die in fuel after only a few hours and thus very few fuel samples contain living bacteria when tested even though the water-bottom may be grossly infected. The dead organisms still plug filters and promote surfactancy. Yeasts survive rather better than bacteria, and fungi, particularly *Cladosporium resinae*, remain viable for many weeks.

It is common to find few organisms alive at the top of an infected fuel tank but numbers increase with depth, being greatest near the fuel/water interface. Thus both the exact location of the sampling point, the degree of turbulence and the age of the sample influence the numbers and types of *living* organisms detected by membrane filtration in the laboratory. Hence interpretation of a result must give regard to all of these factors and they render attempts to impose standards of maximum permissible *viable* counts almost impossible. Nevertheless, Cabral (1980) has proposed a standard of less than 500 fungi per litre. To some extent total biological contamination of fuel, dead or alive, can be assessed by chemical methods, such as protein assays, and total particulate counts on membranes may be valuable.

The declared policy of the Royal Navy is to count biological particles on membranes and if this exeeds 500 per litre the necessity for tank cleaning is indicated.

To illustrate the effects of dispersion and survival Table 1 gives the results from tests on c.500 ml samples of gas oil, each of which had a little associated free water and was several days old at the time of testing.

Table 1   Number of viable micro-organisms/ml

Fuel sample	Water phase			Fuel phase		
	Bacteria	Yeasts	Fungal units	Bacteria	Yeasts	Fungal units
M1	42,000	1,200	100	2	51	10
M2	1,740	2,750	50	0	26	0
M3	17,000	7,900	2,350	1	122	1
L1 (CO)	290,000	150	0	0	66	1
A8	40,000	6,400	0	0	7	0
A9	1,600	300	50	1	3	1

Samples M3 and A8 contained sulphate-reducing bacteria in the water phase but these were not detected in the fuel phase.

A further illustration of these factors is provided by a laboratory experiment in which 200 ml of gas oil with 20 ml of heavily infected water bottom was shaken and left static at room temperature. The water bottom contained initially 105,000,000 viable micro-organisms/ml and these were a mixture of bacteria (including SRB) yeasts and fungi. After 1 day a sample of fuel extracted from 1 cm above the interface had 168 viable micro-organisms/ml and 14 days later a similar sample contained only 14 viable micro-organisms/ml although the water bottom still contained 39,000,000 viable micro-organisms/ml. The residual survivors in the fuel phase were yeasts.

Thus tests for *living* organisms in fuel may yield misleading low results and fail to indicate a very heavy water bottom infection.

The results quoted emphasize the impractability of establishing go/no go standards for living organisms in the fuel phase, as far as the use of that particular fuel load is concerned. However, it must be conceded that fuel containing only a few living organisms is unlikely to infect a different fuel installation if it is transferred to it.

Conventional microbiological tests involve filtering fuel through a membrane which is then washed, transferred to a growth medium and colonies of microbes counted after incubation. Some fuels, because of their viscosity or particulate load cannot be membrane filtered. They can however be emulsified with sterile water and a non-toxic surfactant and are then amenable to standard viable count procedures (Hill, 1975).

ON-SITE TESTING

A much used practical device for on-site microbiological testing for aerobic bacteria, yeasts and fungi, is the Dip-slide (e.g. Easicult, Orion Diagnostic, Helsinki). Its nature and use has been previously described (Hill, 1975; Genner, 1976 etc). In essence a small plastic paddle coated with a nutritive gel is supplied in a sterile container. The paddle is dipped into a sample, drained, then returned to its container and incubated.

One type of gel detects and estimates bacteria within 48 hours and another detects yeasts and fungi within 3–5 days. Another type of test, Sig-Sulphide (Echa Products, Cardiff) rapidly detects anaerobic sulphate-reducing bacteria (Hill, 1984). A one hour colour test, which detects phosphatase enzymes derived from micro-organisms has also been proposed (Hill, 1970) but this is no longer commercially available. All of these tests should be used on the aqueous phase and can be very misleading if used on the fuel phase. It is suggested that on-site tests such as these are conducted at weekly intervals on fuel oil tanks at risk and the results used to indicate the need for further investigation or to trigger an anti-microbial regime.

ANTI-MICROBIAL MEASURES

Fuel Formation

There are three aspects to anti-microbial measures. The first is to so formulate a fuel oil that it is no longer of a suitable nutrient composition for micro-organisms. It can be said that straight chain hydrocarbons are more readily attacked than branched chains and that paraffinic structures are more susceptible than aromatic structures. Little is known at the moment in practical terms of the influence of crude oil sources or different refinery processes but these are now believed to be of prime importance. It can also be said that some additives stimulate growth whilst others, at the right concentration, are inhibitory. However, as far as we know this fundamental aspect of susceptibility to microbial spoilage has not been the subject of comprehensive study, and therefore preventive measures via changes in the fuel composition are not presently feasible.

Good Housekeeping

The second anti-microbial aspect is to avoid microbial problems by preventive housekeeping measures and the third is to actively combat microbial problems as they arise. Both these aspects can be considered at the refinery through the distribution system and in the ship, vehicle or static service tank.

It is axiomatic that minimizing the water phase will also minimize microbial problems as organisms can only grow in a water phase. This is a reasonably practical measure in a ship's settling tank or a road tanker where water drains can be operated daily. In large refinery or distribution depot storage tanks there is often only one drain point at one side and operating this frequently may entail considerable fuel wastage as the water will only move slowly across the bottom of the tank to the drain point. Some water often remains trapped in depressions in the tank bottom. Nevertheless, there will then be a very considerable reduction in microbial numbers in tank water-bottoms as the reproduction rate of the organisms in residual water will not be able to cope with their repeated wastage from the tank. Any measures to prevent water accumulation in pipes or tanks is bound to be beneficial. Rust, scale and surfactants all stimulate microbial growth. In the refinery it may be possible to store products or product components at temperatures which discourage organisms. The range 25°C–35°C is to be avoided. Contamination of clean fuel could be by airborne organisms but it is much more likely to be by contact with fouled pipe-work, dirty tanks, leaks from floating roof seals etc.

Physical Decontamination

Both filtration and centrifugation remove micro-organisms and hence are likely to be beneficial. Whilst exposure to ultraviolet, ultrasound and Gamma rays kills microbes there is no conceivable large-scale method of utilizing these irradiations for fuel. Pasteurization of fuel has been described by Wycislik and Allsopp (1982) but there is no current application of this. All physical methods decontaminate at the point of application only and there is no downstream effect to resist re-growth or re-infection. For that we must turn to chemical methods.

Chemical Decontamination and Preservation

It is important to distinguish between decontamination of infected systems and prevention of infection in clean systems. For the former we must have a fast acting, penetrative chemical; these properties are not so important when chemically preventing microbial growth in clean systems. Many anti-microbial chemicals have been screened for potential use in fuel (Rogers and Kaplan, 1968, and others), and chromates, glycol ethers and organo-borates have been used extensively.

Chromates are water soluble, not fuel soluble; they have been used in tank washes, directly in water bottoms and they have been formulated into tablets for distribution in the bottoms of aircraft tanks. They suffer the problem of all other biocides which are only water soluble in that they can be entrained as concentrated 'slugs' in the fuel and can then cause engine malfunction or corrosion and may pose a severe toxicity hazard during fuel handling.

Ethylene glycol mono-methyl ether and diethylene glycol monomethyl ether are both fuel and water soluble. They are only moderately anti-microbial but are widely used in aircraft fuel at concentrations of 0.1%–0.15%. Although they are fuel additives they are preferentially water soluble and hence will diffuse from fuel into any water phase with which it is in contact. However, they are not suitable for general use as their prior purpose in aircraft fuel is as an anti-freeze and their high in-use concentration can only be justified as a dual purpose additive. The glycol ethers actually stimulate microbial growth at low concentrations and hence they cannot be used intermittently, which is another major disadvantage for general use. The organo-borate formulation Biobor JF (Consolidated Borax Inc) has been available for thirty years as an anti-microbial fuel additive. Unlike the glycol ethers, sub-lethal concentrations are not stimulatory and 'shock' doses of 135 ppm or 270 ppm in fuel can be used intermittently. However, even the higher dose takes several days to kill microbes in associated water and hence Biobor JF is not suitable for rapidly decontaminating fuel systems. Biobor JF is not completely combustible and most engine builders only approve Biobor JF for intermittent use. There are obvious advantages in using biocides with both fuel and water solubility as whether or not water is present the biocide will disperse and not remain as a 'slug' and will therefore be less of a toxicity or engine hazard; also all free water present will take up biocide until equilibrium concentrations are established. As a rule of thumb, if the volume ratio of fuel:water is less than 400:1, higher concentrations of Biobor JF or glycol ethers will be necessary in order to establish an adequate anti-microbial concentration in the water phase.

Benomyl is used in fuel at a few ppm but it has anti-fungal activity only and bacteria are unaffected. Several biocides (Grotan OX, Bodoxin, Omadine TBAO, Bioban FP, TBHP, Karacide P etc) can all be considered for fuel use. As yet there is incomplete data on solubility and possible adverse affects such as corrosiveness, surface activity and affect on fuel flash point but they are generally much faster in action than the previous chemicals and hence are more suitable for decontaminating dirty systems. Dead organisms do not disappear and in fact they tend to be released from pipe-work and tank-sides; the first consequence of biocide treatment may therefore be an increase in suspended matter in the fuel. Manual cleaning and de-sludging before biocide use is therefore advisable. There is no single suitable concentration of any biocide. The correct concentration is dependent on the fuel:water ratio, the level of contamination, the types of organism present, the temperature and the time available and laboratory tests may be necessary to determine this. If the appropriate concentration proves to be high, the treated fuel may need to be diluted with clean fuel before use. In a ship's settling tank we are unlikely to have more than 24 hours available to clean and sterilize the tank and associated pipe-work but at least the system is relatively small and accessible.

In a large tank we may have restricted access and no method of applying the biocide other than as a spray from the top. Thus, the fuel depth should be reduced to say 0.5 m both to reduce the amount of biocide needed and to give the biocide a fair chance to permeate the fuel and contact any free water phase. However, the fuel:water ratio may then be only 20:1 and the biocide concentration applied must be selected accordingly.

Whilst laboratory experiments or data sheets can indicate the concentration of a biocide which will disinfect a water-bottom sample in a given time it is often difficult to predict the concentration of biocide which will actually diffuse into a tank water bottom or its time of arrival, if the application of the chemical is to the bulk fuel or the fuel surface. However, a simple device now exists (Hill, 1986) for determining the concentration of any biocide in water and hence during an actual disinfection regime water samples can be taken and assayed for biocide.

Thus adjustments to the disinfection regime can subsequently be made to ensure that target concentrations of biocide are being achieved for an adequate time. The device (Echa Biocide Monitor, Echa Products, Cardiff) consists of a plastic strip carrying at one end a small pad impregnated with dried nutrients, spores of an organism which is very sensitive to biocides and a growth indicating dye. The strip is dipped into the sample for a few seconds, drained, placed in an individual plastic incubating chamber, and held at 35°C overnight. An effective anti-microbial chemical in the water sample will prevent growth of the detector organism and the pad will remain white. Ineffective concentrations of biocide will permit growth and the pad turns red. The device can be calibrated against known concentrations of the biocide in use. Thus, by testing a range of dilutions of the water samples taken the actual concentration of biocide present can be determined.

Microbial slimes tend to adhere to pipes but dismantling and cleaning is rarely justified. Circulation of biocide through *full* pipes is usually acceptable.

In some fuel tanks water displaces the fuel so as to maintain a constant trim situation. In this case direct injection of biocide into the water phase may be necessary for biological control although there are then problems of environmental pollution to be resolved when the water phase is discharged.

The biological problem in very large long-term fuel stores is somewhat different. As oxygen is not freely available, aerobic growth of organisms may be restricted and the potential hazard may be from sulphide produced by anaerobic sulphate-reducing bacteria. A specific anti-microbial chemical targeted against SRB is now available ('Minder', May & Baker Ltd, Dagenham, UK) and its mode of action against SRB has been recently published (Hill, 1986).

CONCLUSION

Microbial fouling in distillate fuel is a problem of our time and is likely to continue. It can be minimized by good housekeeping. A range of anti-microbial chemicals is available; their method of use and the concentration applied must be carefully planned and their effectiveness assessed by on-site or laboratory microbiological tests.

REFERENCES

Arnold, J. B. (1983). 'The effects of microbial growth and its by-products on the coalescing/filtration of hydrocarbon fuels', *Proceedings of Filtech Conference, UK 1983*, pp. 51–61.

Cabral, D. (1980). 'Corrosion by microorganisms of jet aircraft integral fuel tanks. I. Analysis of fungal contamination', *Int. Biodet. Bull*, **16**, 23.

Genner, C. (1976). 'Evaluation of the "Dip-slide" technique for microbiological testing of industrial fluids', *Proc. Biochem.*, **11**, 39.

Hill, E. C. (1970). 'A simple microbiological test for aircraft fuel', *Aircraft Engineering*, July 1970, pp. 24–28.

Hill, E. C. (1975). 'Biodeterioration of petroleum products', in *Microbial Aspects of the Biodeterioration of Materials*, (eds. R. J. Gilbert and D. W. Lovelock), pp. 127–136. Applied Science Publishers Ltd, London.

Hill, E. C. (1983). 'Microbial aspects of corrosion, equipment malfunction and systems failure in the marine industry', *General Council of British Shipping Rept. TR/104*, pp. 1–65.

Hill, E. C. (1984). 'a. Microorganisms — number, types, significance, detection', in *Monitoring and Maintenance of Aqueous Metal-working Fluids* (eds. K. W. A. Chater and E. C. Hill), pp. 97–112, John Wiley & Sons, Chichester, UK.

Hill, E. C. (1984). 'b. Biodegradation of petroleum products', in *Petroleum Microbiology*, (ed. R. M. Atlas), pp. 579–617, Macmillan Publishing Co., New York.

Hill, E. C. (1986). 'Sulphide Generation in Metal-working fluids and its control', *Proc. 'Additives for Lubricants and Operational Fluids'*, Volume II, 11–10, Technische Akademie Esslingen.

Rogers, M. R. and Kaplan, A. M. (1964). 'A survey of the microbiological contamination in a military fuel distribution system', *Dev. Ind. Microbiol.* **6**, 80.

Shennan, J. L. 'A recommended method for testing distillate fuels', *Society of Applied Bacteriology Technical Series No. 23*, (in press).

Thaysen, A. C. (1939). 'On the gas evolution in petrol storage tanks caused by the activity of microorganisms', *J. Inst. Petrol*, **25**, 411.

Unzelman, G. H. (1984). 'Diesel fuel demand — a challenge to quality'. *Institute of Petroleum Technical Paper IP 84 — 001*, pp 49.

Wycislik, E. and Allsopp, D. (1983). 'Heat control of microbial Colonisation of shipboard fuel systems', in *Biodeterioration 5*, (eds. T. A. Oxley and S. Barry), pp. 453–461, John Wiley & Sons, Chichester, UK.

This paper is an extended version of an article which was published in the December 1985 issue of the Marine Engineers Review and the author acknowledges the editor's consent for using this material.

BACTERIAL BIOFILM DEVELOPMENT AND ITS EFFECTS ON MICROBIAL PROCESSES IN OIL RECOVERY SYSTEMS

J. W. Costerton

AOSTRA Professor of Microbiology
University of Calgary

In all natural and industrial aquatic environments, bacteria show a pronounced tendency to adhere to available surfaces and to proliferate to form matrix-enclosed biofilms. Because of this sessile mode of growth, bacteria within these coherent plaque-like accretions are often undetected by conventional sampling techniques, but it is these biofilm bacteria that are responsible for much of the corrosion, injection well plugging, and reservoir souring that affects oil recovery operations. We have developed a series of biofilm samplers for use in various locations, and at various pressures, that allow us accurately to monitor the development of bacterial biofilms on pipe and tank surfaces, and we have developed systems in which the development of bacterial biofilms on formation rock surfaces can be modelled. All of our experiments to date have shown that bacteria within biofilms are much more difficult to control with biocides than their floating (planktonic) counterparts in these systems, and we submit that biocide efficacy data is not applicable to practical oilfield situations unless it is generated in relation to organisms growing in established biofilms.

*This paper was presented at the meeting but is unavailable for publication

POSTERS

INFLUENCE OF SULPHATE-REDUCING BACTERIA
ON ELECTRON FLOW BETWEEN TWO STEEL ELECTRODES

Sylvie Daumas[1,2], Ralf Cord-Ruwisch[1], Jaqueline Crousier[2]

[1]*Laboratoire de Microbiologie ORSTOM,*
[2]*Laboratoire de Chimie des Materiaux, Université de
Provence, 3 Place Victor Hugo, F-13331 Marseille Cedex 3*

In recent literature three possibly cumulative mechanisms are supposed to be responsible for anaerobic corrosion of steel caused by the activity of sulphate-reducing bacteria (SRB):

1. Cathodic depolarization by bacterial oxidation of hydrogen from the polarized metal surface.
2. Stimulation of the anodic dissolution process by the produced sulphide $(Fe^{2+} + HS^- \rightarrow FeS\downarrow + H^+)$.
3. Favouring of the cathodic reaction $(2H^+ + 2e^- \rightarrow H_2)$ by precipitates of FeS, which are cathodic to iron.

The aim of the present work was to separate the influences of these three processes. An electrochemical cell consisting of two identical half-cells with steel electrodes was used to measure the electron motive force or the current between both electrodes during the growth of SRB in only one of the half cells.

The results supported the classical theory of anaerobic corrosion and proved that the oxidation of cathodically formed hydrogen was the major corrosive mechanism. Furthermore, this mechanism has been shown to occur only in the direct vicinity (≤ 1 cm) of the steel surface, probably by direct contact of the bacterial cells with the metal, and not via diffusion of hydrogen through the liquid phase. These conclusions originated from the following observations.

1. The surface area in contact with SRB became cathodic during the growth of hydrogen oxidizing SRB. An electron-motive force of 30 mV established between both identical steel electrodes.
2. The growth activity of hydrogen consuming SRB (*Desulfovibrio vulgaris*) caused a corroding electron flow (current density of 2 to 6 $\mu A.cm^2$)

from the sterile surface to the surface in contact with the bacteria. After the growth of the bacteria, only the electrode from the sterile half-cell was visibly corroded.

3. The addition of dissolved sulphide to the sterile (anodic) surface area did not additionally stimulate the corroding current.

4. During the growth of SRB unable to oxidize hydrogen (*Desulfovibrio sapovorans*) no current developed.

5. The growth activity of hydrogen consuming SRB, which did not produce sulphide (fumarate replaced sulphate as electron acceptor), also created a corroding electron flow attacking only the sterile steel surface. However, this current was somewhat less intense than the current observed in sulphide-producing cultures (see 2). This may be due to the absence of FeS as a cathodic stimulant or to the slower growth of the bacterial strain with fumarate as electron acceptor.

6. Separation of the SRB from the steel surface by means of a dialysis membrane (minimal distance to the metal surface: 1 cm), prevented the utilization of cathodic hydrogen.

CORROSION OF STEEL BY ANAEROBIC BACTERIA, THEIR PRODUCTS AND MIXED COMPONENT BIOFILMS

Nicholas J. E. Dowling[1], Jean Guezennec[2] and David C. White[3]

[1]*Department of Biological Science, Florida State University, Tallahassee, Florida 32306 USA,* [2]*IFREMER Centre de Brest, BP 337, Brest Cedex, France,* [3]*Institute for Applied Microbiology, University of Tennessee, Knoxville TN 37996, USA*

Corrosion of metal alloys in the environment appears to be related at least in part to the presence of anaerobic bacteria held within biofilms. Sulphate-reducing bacteria, which have been implicated in such corrosion, are obligate anaerobes which may pass through the 'aerobic' zone to reach niches which are anaerobic by virtue of the activities of oxygen-utilizing bacteria. Sulphate-reducing bacteria are easily detected in biofilms using fatty acid biomarkers. These results can be confirmed by cultural methods. *Vibrio natriegens* can provide carbon substrate and an anaerobic environment for growth by *Desulfobacter postgatei* 2ac9.

'Signature' biomarker lipid fatty acids of *Desulfobacter* have been detected in thin biofilms on titanium pipes exposed to aerobic rapidly flowing seawater at IFREMER in Brest, France. These studies show that sulphate-reducing bacteria can be transported to colonize and function in aerobic systems.

MARINE FOULING AND CORROSION STUDIES ON 90:10 (1.5% FE) CUPRO-NICKEL ALLOY

B. J. Garner, A. H. L. Chamberlain and J. E. Castle

University of Surrey, Guildford, Surrey, GU2 5XH

Cupro-nickel alloys, especially 90:10 (1.5%Fe), are being used successfully in a diversity of immersion applications such as boat hulls, fish-cage meshes and splash-zone sheathing of offshore platforms. The alloy is unique in offering both corrosion and fouling resistance although temporary microfouling may occur in certain sheltered environments.

The sequence of molecular and biological fouling and corrosion events occurring in both a sheltered harbour system with high biological productivity and in laboratory systems have been monitored using a range of microbiological, surface analytical and electro-chemical techniques.

X-ray photoelectron spectroscopy revealed a rapid adsorption of dissolved marine organic materials which became incorporated into the thin surface film of corrosion product. As the exposure time increased there was a change in the composition of the organic film with more oxidized carbon compounds and more nitrogen-rich materials adsorbing.

After an 'ageing' period of at least four weeks in Chichester Harbour the alloy became colonized by bacteria and diatoms which modified the surface toxicity allowing organisms with less copper tolerance to colonize. Sporadic and annual exfoliation of either part or the whole of the fouling community was observed, resulting in a self-cleaning effect. Electrochemical investigations indicated that microfouled cupro-nickel subjected to a removal of oxygen lost its normal cathodic inhibition and corrosion could ensue, at least temporarily, when oxygen was replenished.

THE ROLE OF SULPHATE-REDUCING BACTERIA IN HYDROGEN ABSORPTION BY STEEL

C. H. J. Parker, K. J. Seal and M. J. Robinson

Biotechnology Centre & School of Industrial Science, Cranfield Institute of Technology, Cranfield, Beds.

Biogenic sulphide production in marine environments is known to encourage the embrittlement of steels due to an increased susceptibility to hydrogen absorption. However, it remains unclear to what extent the sustained presence of sulphate-reducing bacteria (SRB) is necessary for enhanced hydrogen absorption to occur. By measuring the permeation of hydrogen through steel membranes electrochemically, it is possible to measure levels of hydrogen absorption for varying sulphide concentrations, using different SRB species. Initial results indicate the absorption by steel of less than 2% of total cathodic hydrogen produced during corrosion and the establishment of a surface hydrogen concentration in steel of less than 0.02 ppm. Hydrogen permeation peaks, coinciding with increasing sulphide production and sulphide film formation, have been produced within 2 days of inoculation with a SRB culture. The influence of sulphide concentration and surface film growth/composition is of importance, with respect to hydrogen permeation through steel membranes. The role of bacterial adhesion and its possible influence on hydrogen absorption is also involved.

PHYSIOLOGICAL ASPECTS OF SULPHATE-REDUCING BACTERIA INVOLVED IN ANAEROBIC CORROSION

C. Chatelus, M. Czechowski, M. F. Libert-Coquempot, R. Toci, A. Euve, G. Fauque and J. Le Gall

ARBS, Section Enzymologie et Biochimie Bactérienne, CEN Cadarache, 13108 Saint Paul Lez Durance Cedex, France

The anaerobic corrosion of metal involves several types of micro-organisms such as sulphate-reducing bacteria, methanogenic bacteria (archaebacteria), sulphur-reducing bacteria (*Desulfuromonas*) and photosynthetic bacteria (*Chlorobium, Chromatium*).

Of particular interest are the physiological aspects of sulphate-reducing bacteria under conditions where corrosion might occur. Hydrogen consumption and sulphate reduction contribute to this corrosion process by means of hydrogenase and sulphate reductase, the two key enzymes involved in corrosion.

In contrast with other published results, a hydrogenase activity has also been found within the genus of sulphate-reducing bacteria *Desulfotomaculum* indicating the possible role of sporulated micro-organisms in corrosion phenomena (Lespinat *et al.*, 1985).

Recently three proteins (hydrogenase, desulfoviridin and a molybdoprotein) have been purified and characterized from *Desulfovibrio salexigens*, the only well-known halophilic strain of sulphate reducer (Czechowski *et al.*, in press).

Growth curves have been established for some sulphate-reducing strains of the genus *Desulfovibrio* at different temperatures. A mesophilic (optimal growth temperature around 35°C) *Desulfovibrio baculatus* strain 9974 can grow at temperatures near 10°C indicating its adaption capacity to climatic variations and its possible role in corrosion occurring in cold seawater.

REFERENCES

Lespinat, P. A., Denariaz, G., Fauque, G., Toci, R., Berlier, Y. and Le Gall, J. *C.R. Acad. Sci.*, (1985), **301**, 707–710.

Czechowski, M., Fauque, G., Galliano, N., Dimon, B., Moura, I., Moura, J. J. G., Xavier, A. V. and Le Gall, J. (1986). *J. Indust. Microbiol.*, (in press).

RESISTANCE OF CONCRETE TO
MICROBIAL SULPHURIC ACID CORROSION

Wolfgang Sand and Eberhard Bock

*Mikrobiologie, Universität Hamburg,
Ohnhorststr. 18, D-2000 Hamburg 52, FRG*

A microbiological investigation of the microflora growing on the concrete walls of partially filled trunking sections, showed that sulphuric acid producing bacteria of the genus *Thiobacillus* were responsible for the rapid degradation of the concrete above the water level. The species *Thiobacillus* neapolitanus, T. intermedius, T. novellus, and *T. thio oxidans* were isolated. *T. thio oxidans* is the indicator for severe corrosion.

Based on these results biogenic sulphuric acid corrosion was reproduced in the laboratory by means of an incubation chamber with a strictly controlled atmosphere. In this chamber the conditions were optimized for thiobacilli: 30°C, more than 98% relative humidity and a reduced sulphur compound as substrate. Thirty-two different concrete samples were inoculated with 10^{13} cells of the 4 thiobacillus species which had been isolated from trunking section walls and incubated for 9 months. The tests were run with hydrogen sulphide, thiosulphate and methylmercaptan. Corrosion was demonstrated only with the first 2 compounds.

After 9 months of incubation at a hydrogen sulphide concentration of $10^{\pm}1$ ppm, severe corrosion was noted with the concrete test blocks. A medium weight loss of 3.5% with values running up to 10% were demonstrated. The pH-value in the surface water had decreased to pH 1 and below. Up to 10^8 cells of *T. thio oxidans* per cm^2 were measurable. All these results were reproduced in a replicate experiment. Significant differences were noted for several cement types. As a rule blast-furnace-type cement yielded poor results, whereas Portland-type cement had good resistance to biogenic attack.

After 9 months of incubation with thiosulphate, corrosion was also noted with the concrete test blocks. A medium weight loss of 1.8% was demonstrated. The pH-values decreased to a range of pH 2 and 3. Up to 10^7 cells per cm^2 *T. thio oxidans* were measured on the concrete test blocks.

With methylmercaptan, which was taken as a representative of the organic volatile sulphur compounds in sewer atmosphere, corrosion was not measurable. In 2 experiments at concentrations of 10 and 1 ppm methylmercaptan growth of thiobacilli could not be demonstrated. Therefore, these compounds can be disregarded as a directly metabolizable sulphur source for thiobacilli.

OPTIMIZATION OF BACTERIAL GROWTH MEDIA IN OFFSHORE WATER SYSTEMS

A. V. Maxwell[1], S. Maxwell[1,2] and K. M. McLean[2]

[1]*Media Supplies, Devanha House,*
Riverside Drive, Aberdeen [2]*Corrosion Specialists*
(North Sea) Ltd, Devanha House, Riverside Drive, Aberdeen

The growth of bacteria in offshore water systems has been implicated in pitting corrosion of process equipment. The use of liquid broth medium together with serial dilution techniques is the most common means of enumerating these bacteria. In most cases bacteria, and in particular sulphate-reducing bacteria (SRB), are enumerated only in the bulk phase; normally being counted in single or duplicate dilution series. The following points, now widely accepted in the offshore industry, clearly show that this type of monitoring allows only minimal information to be gathered:

(a) It has been clearly demonstrated that, in these systems, only those bacteria attached to metal surfaces influence corrosion.
(b) Only by performing triplicate dilution series can reliable bacterial counts be obtained and only this number of replicates allows order of magnitude differences in bacterial numbers be determined with reliable significance.
(c) The media generally employed for counting marine bacteria are normally only 1–10% efficient at recovering viable bacteria from a sample.

Bearing these points in mind, use of growth media to assess bacterial problems can only be optimized by including in the monitoring programme a screening test to determine the medium showing the most efficient recovery, performing at least triplicate dilution series and, most importantly, regularly assessing the degree of bacterial fouling on internal metal surfaces.

AN ASSESSMENT OF VIABLE COUNT PROCEDURES FOR ENUMERATING SULPHATE-REDUCING BACTERIA WITHIN AN ESTUARINE SEDIMENT WITH HIGH RATES OF SULPHATE-REDUCTION

G. R. Gibson, R. J. Parkes and R. A. Herbert

Scottish Marine Biological Association, P.O. Box 3, Oban, Argyll and Department of Biological Sciences, University of Dundee

Sulphate reduction is an important process in the terminal stages of organic matter mineralization in marine and estuarine environments. Anaerobic sulphate reduction has also economic and industrial importance, especially in the oil industry where growth of sulphate-reducing bacteria can clog equipment, sour oil and gas, corrode pipelines and produce toxic hydrogen sulphide gas. Although there exist accurate procedures for estimating *in situ* rates of sulphate reduction, methods for estimating the biomass and composition of *in situ* populations of sulphate-reducing bacteria are not so well developed.

A technique which is often used to enumerate populations of sulphate-reducing bacteria is the viable count procedure. However, few studies have been undertaken to determine the relative efficiencies of the various media used to enumerate the different types of sulphate-reducing bacteria. We have determined the recovery of known viable populations of sulphate-reducing bacteria belonging to the genera *Desulfovibrio, Desulfobulbus* and *Desulfobacter* from autoclaved, anoxic estuarine sediments using Postgate's and Widdel's medium. Recovery of *Desulfovibrio* populations was consistently higher with Postgate's medium, whilst Widdel's medium always yielded higher viable counts of *Desulfobulbus* and *Desulfobacter*, but in all cases average recovery after 5 days' incubation at 4°C was only about 50% of the added viable cells. Two surface active agents, cetyl trimethylammonium bromide (CTAB, final concentration 0.00001 % w/v) and sodium tripolyphosphate (NaPP, final concentration 0.0005 % w/v) both increased recoveries of viable cells, with CTAB showing the greatest effect (17% increase). The addition of CTAB to untreated sediment samples also significantly increased the viable counts of *in situ* populations of sulphate-reducing bacteria although the effect was somewhat more variable (0–200% increase). Viable counts of different types of sulphate-reducing bacteria from estuarine sediments, even using the most appropriate medium and in the presence of CTAB, grossly underestimate the *in situ* populations of sulphate-reducing bacteria when compared with direct bacterial counts and rates of sulphate reduction measured within the same sediment. The significance of viable counts of sulphate-reducing bacteria within estuarine sediments should therefore be interpreted with caution.

THE USE OF SEROLOGICAL TECHNIQUES FOR THE DETECTION OF SULPHATE-REDUCING BACTERIA IN THE OIL INDUSTRY

S. Bobowski

British Petroleum PLC, Research Centre, Sunbury-on-Thames

Sulphate-reducing bacteria (SRB) comprise a group of morphologically diverse, obligately anaerobic organisms which obtain their energy by dissimilatory reduction of sulphate to sulphide. The SRB are important for a number of economic and ecological reasons. In relation to the oil industry they can be found as contaminants of drilling fluids used in oil exploration and production operations, as part of the microbial population of the injection seawater used in secondary oil recovery and, if injection water systems are untreated, as *in situ* inhabitants of the water base of producing fields. Their activities cause corrosion of pumping machinery and storage tanks and pollute oil products by introducing reduced sulphur into oil and its associated gas. Their economic effects are therefore of serious concern to the oil industry.

Any ecological study of SRB requires their identification in mixed microbial communities. Traditional methods of enumeration and classification are very time-consuming but immunological techniques provide one way to overcome these problems.

Serological techniques such as agglutination, immunodiffusion, immuno-electrophoresis + enzyme linked immunosorbent assay (ELISA) can be used for the identification and classification of SRB. A method has been developed for 'antigenic fingerprinting' of SRB with polyclonal antibody probes, as well as a 'designer' ELISA method for detecting low numbers (as low as 10 cells per filter) of SRB in culture or natural samples. An indirect micro-ELISA technique employs unique combinations of polyclonal antiserum specifically designed for environments of interest such as the North Sea. This procedure is quick, simple and efficient with respect to time, reagents and equipment, and of high sensitivity in comparison to the traditional MPN counts.

The development of SRB monoclonal antibody probes is a further possibility for specific direct detection of SRB *in situ*.

RAPID TECHNIQUES FOR THE DETECTION AND QUANTIFICATION OF SULPHATE-REDUCING BACTERIA

P. E. Cook and C. C. Gaylarde

Department of Biological Sciences, City of London Polytechnic, Old Castle Street, London E1 7NT

The sulphate-reducing bacteria (SRB) are a group of anaerobic, heterotrophic organisms which are responsible for a number of the problems experienced in the offshore oil industry. Their peculiar metabolism, involving the utilization of inorganic sulphate as a final electron acceptor, leads to the production of large amounts of hydrogen sulphide, causing the souring of crude oil and health hazards on the platforms. The production in anaerobic areas of ferrous sulphides may induce metal corrosion which can lead to failure of the platform structure or pipework and to blockages of oil-bearing strata, filters, valves, etc.

Reliable methods for the detection and enumeration of SRB are few and mostly involve prolonged incubation periods in specialized media and the attainment of anaerobic conditions.

A number of methods developed at the City of London Polytechnic have the advantages of speed and/or simplicity compared with the commonly used enumeration technique, the most probable number (MPN) method.

The new techniques are an ELISA method using specific antisera, an MPN technique modified to use microtitre plates as the incubation vessels and an agar column method which allows enumeration by the measurement of depth of blackening within the medium. The standard MPN method (API RP38) has recently been critically reviewed by a task group of the National Association of Corrosion Engineers.

The ELISA method is rapid, extremely sensitive and amenable to high replication. It has the disadvantages, however, of requiring a trained operative and specialist facilities. Additionally, the technique measures total cell numbers and does not differentiate between active and inactive bacteria.

The microplate MPN method is economical of both materials and operator time. It requires a degree of manipulative ability, but no specialist personnel. Because of the low cost and speed of the technique, multiple replication can lead to a very high level of sensitivity.

The agar column method is extremely simple and requires no specialist facilities apart from an autoclave or pressure cooker and some means of maintaining temperature. No dilution steps are necessary. The sensitivity is lower than for the other two methods, although this can be increased by the addition of iron to the test system.

The three methods would find uses in different situations.

MECHANISMS OF MICROBIAL
TRANSPORT THROUGH POROUS ROCKS

F. K. Dow and T. M. Quigley

British Petroleum PLC, Research Centre, Sunbury-on-Thames

Degradation of subsurface petroleum accumulations, which is widely believed to result from bacterial oxidation, can adversely affect the economic worth of a petroleum prospect. The extent to which petroleum is biodegraded within sedimentary rocks must then depend upon the potential activity of the resident microbial population, controlled by factors such as temperature and nutrient availability. Of equal importance, however, is the influence of the porous rock media to which bacteria must gain access. Although the former controls on biodegradation have received considerable attention, the rates and mechanisms of bacterial growth and transport in porous rock are poorly understood.

In the present study a system of sterile nutrient reservoirs connected by static nutrient-saturated reservoir-type rock plugs were used to determine the rates of bacterial growth and transport through porous rock. A series of experiments were performed varying growth rates and core dimensions. The resultant data were interpreted using a mathematical model which accurately described bacterial growth kinetics and transport by diffusion.

It was found that the bacterial penetration rate depended upon both the *growth rate*, a function of incubation temperature in a constant nutrient regime, and the *diffusion rate*, which was defined by the bacterial cell/rock system. Only by clearly separating and modelling these two effects can meaningful extrapolations from laboratory to subsurface conditions be made.

Diffusion coefficients within the range 10^{-12} to $10^{-11}\,\mathrm{m^2\,s^{-1}}$ for *E. coli* through Berea sandstone were determined. These values indicate that under subsurface conditions (ie, limited growth conditions) bacteria can diffuse through reservoir rocks (mean pore size greater than *ca* 20 μm) over tens of metres in geologically reasonable times (i.e. less than *ca* 1–10 Million Years). Bacteria do not diffuse through fine grained rock such as compacted shales.

A STUDY OF THE MICROBIAL DECOMPOSITION OF OIL UNDER ANAEROBIC CONDITIONS USING GEL-STABILIZED MODEL SYSTEM

Elaine A. Rees, Caroline S. Tughan,
Moya A. Russ and R. A. Herbert

Department of Biological Sciences, The University, Dundee

Sulphate-reducing bacteria cause major economic and operational problems in the oil industry and are frequently isolated during oil recovery operations. *Desulfovibrio* spp have a restricted carbon metabolism and there is still no convincing experimental evidence available to show unequivocally that these bacteria can utilize hydrocarbons as carbon and energy sources. It has been proposed that *Desulfovibrio* spp causing problems in the oil industry utilize metabolic end-products generated by primary hydrocarbon utilizers. In this study biphasic gel-stabilized laboratory models were used to investigate the anaerobic decomposition of North Sea oil using defined mixed microbial populations of hydrocarbon utilizers and sulphate-reducing bacteria. The gels were established in beakers and consisted of a semi-solid agar basal layer containing a mineral salt medium and the microbial inoculum. When the basal layer had gelled under an inert gas atmosphere it was overlain by the oil layer. The gels were incubated under anaerobic conditions at 23°C and at time intervals the physico-chemical gradients which developed (pH, Eh and soluble S^{2-}) were determined using needle-electrodes mounted on a micromanipulator. The spatial distribution of bacterial populations which developed were determined by aseptically removing vertical gel cores (10 mm diameter) which were then sliced into 2.5 mm thick sections.

Viable bacterial populations in each depth horizon were determined using agar shake culture techniques. The gel model system described offers considerable promise as a research tool for the investigation of the microbial processes involved in the decomposition of North Sea oil under laboratory conditions.

THE EFFECTS OF LOW CONTINUOUS
CHLORINATION ON THE MARINE BIOFOULING

J. Guezennec[1], M. Therene[1] and D. C. White[1,2]

[1]*IFREMER Centre de Brest, BP 337,*
29273 BREST CEDEX, [2]*Institute of Applied*
Microbiology, ORNL, Oak Ridge, TENNESSEE

Biofouling formed on the inside surface of heat exchange tubes may cause increase in heat transfer resistance and fluid frictional resistance which results in energy losses. Microbial activity in the biofilm may also initiate pitting corrosion.

Chlorine as hypochlorous salts is the most usual slimicide to prevent biofouling and remove formed slime layers. However, the use of chlorine may lead to the formation of unacceptable pollutants for the marine environment.

Biochemical methods using lipids as 'biomarkers' may give an insight into the microbial community attached to the surfaces. Biofouling has been studied by lipids analysis, microbiological examinations and heat transfer measurement.

Results indicated that low continuous chlorination (0.05 mg/l as total residual chlorine) is not sufficient to remove and destroy all the biofilm at low flow velocity (0.1 m/s). At higher velocity (1.5 m/s) detachment of the biofilm has been achieved by the combined action of chlorine and fluid shear forces.

Based on these results, the reaction and transport processes contributing to the chlorine/biofilm reaction are assumed to be as follows:

(a) diffusion of the biocide from the seawater through the biofilm
(b) chemical reaction with the extracellular polymers leading to the oxidation of some of them and incorporation of chlorine in bacterial cells.
(c) detachment of the biofilm by the fluid shear forces. If not, the residual matter may act as a new substrate for another bacterial colonization.

Chlorine, therefore, even at high flow velocity (1.5 m/s) is not sufficient to remove all the biofilm and better control will be obtained with the combination of mechanical and chemical cleaning methods

MICROBIAL CONTAMINATION OF
IN-USE WATER-BASED DRILLING MUDS

F. K. Dow

British Petroleum PLC, Research Centre, Sunbury-on-Thames

Water-based drilling muds are widely used in land-based drilling operations, exploration drilling and during the initial stages of drilling in deeper offshore wells. Comprising up to 50% suspended or dissolved clays, polymeric materials and other biodegradable organics in saline water, they can provide an ideal substrate for microbial growth. The increasing use of biodegradable polymers, eg modified cellulose, starch and Xanthan, in mud formulations make them particularly sensitive to bacterial attack.

The possible consequences of bacterial growth and activity include an increase in mud corrosivity through production of corrosive metabolites, or a loss in mud condition.

To help assess the extent of potential microbiological problems, mud samples from a number of offshore and land-based drilling operations were examined. The muds were often highly contaminated with aerobic (up to 10^9/g mud), anaerobic (up to 10^8/g mud) and sulphate-reducing bacteria (up to 10^4/g mud). High pH (up to 12.5) mud types (clay or polymer) and moderate salinities (up to 3% wt/wt chloride) had no apparent detrimental effect on microbial numbers. However, only very low levels were detected in salt-saturated muds. Efficient biocide treatment also reduced numbers to low or undetectable levels.

Yeasts and moulds were not detected. The highly saline, alkaline conditions common to most water-based muds appeared to inhibit their growth.

As large viable populations of bacteria could have a deleterious effect on mud condition, an assessment of the efficiency of current biocide application and monitoring practices is required in addition to improvements in the general handling of drilling fluids in the field to minimize microbial contamination.

MICROBIAL ASPECTS ON CRUDE OIL
STORAGE ON A WATER BED IN ROCK CAVERNS

R. Roffey, A. Norqvist and A. Edlund

*Department of Microbiology, National Defence
Research Institute (FOA 4), S-901 82 Umeå, Sweden*

In Sweden a large number of mined rock caverns in granite for long-term storage of different petroleum products and crude oil have been in operation for many years. These unlined caverns are situated underground below the ground water table to ensure a sufficient water pressure on the oil to prevent leakages. For crude oil storage a fixed water bed is used.

When storing jet fuel and heavy fuel oil in this type of storage for long periods of time some microbial problems have, in some cases, been observed (1,2). In order to investigate whether microbial problems could be anticipated in caverns containing crude oil a microbial study was carried out.

Two rock cavern plants for storage of crude oil were chosen for this investigation, one on the Swedish east coast and the other on the west coast. Samples were collected of the inflowing groundwater, the bedwater and the outflowing water and analysed for different microbial groups, microbial activities and a large number of chemical components.

In the two rock cavern plants differences could be seen in the water quality before it entered the caverns and the water already *in situ*. Changes in a number of parameters have been observed.

The levels of free oxygen were lower in the caverns than in the incoming water indicating that oxygen was consumed. Values from 4 to 32 mg O_2/m^3/day were calculated. From the values of biochemical oxygen demand it could be seen that there was biological oxygen consumption indicating that organic matter, in this case, crude oil was consumed.

The amounts of aerobic micro-organisms were lower in the caverns that in the incoming water and the opposite was the case for anaerobic bacteria like sulphate-reducing bacteria, indicating that the environment was oxygen-limited. The anaerobic situation was confirmed by the presence of methane and hydrogen sulphide in the bedwater. Calculations showed that 0 to 83 mg H_2S/m^3/day and 0 to 32 mg/m^3/day of methane was produced per cubic meter of water running through the caverns.

The heterotrophic activity measured as ^{14}C-glutamate converted to CO_2 and the oil degrading activity measured as ^{14}C-hexadecane and

251

^{14}C-naphthalene converted to CO_2 showed that these activities were very low in the caverns. No anaerobic conversion of labelled compounds were observed.

From these studies and others (1–4) it is probable that the microbial degradation of crude oil in this storage system is oxygen limited which implies that it is preferable to decrease the inflow of fresh water into the caverns to minimize the amount of available oxygen.

Due to the low microbial activities observed it is not foreseen that any major problems due to micro-organisms will be found when storing crude oil in this type of rock cavern if suitable precautions are taken when choosing storage sites with preferably fresh water, and carefully evaluating the cavern design.

REFERENCES

Roffey, R., Norqvist, A. and Edlund, A. (1983). 'Microbial problems in connection with long term storage of petroleum products p D 56–72', in *Proceedings from Conference on Long Term Storage Stabilities of Liquid Fuels, Tel Aviv*, (ed. N. Por), Israel Institute of Petroleum and Energy.

Roffey, R., Norqvist, A. and Edlund, A. (1984). 'Biodeterioration of jet fuel during longterm storage in rock caverns', in *Proc. from 6th Int. Biodeter. Symp. Washington Biodeterioration 6*, Commonwealth Agricultural Bureau, Slough (in press).

Ross, D. (1983). 'Ecological studies on sulphate-reducing bacteria in offshore oil storage systems'. *Thesis*, University of Aberdeen.

Wilkinson, T. G. (1983). 'Offshore monitoring', in *Microbial Corrosion*, The Metals Society, London.

MICROBIAL COMMUNITIES IN
OFFSHORE OIL STORAGE SYSTEMS

Keith M. McLean[1] and W. Allan Hamilton

*Department of Microbiology, University of
Aberdeen, Marischal College, Broad Street,
Aberdeen [1]Present Address: Corrosion Specialists
(North Sea) Limited, Devanha House, Riverside Drive, Aberdeen*

As a direct consequence of the ability of sulphate-reducing bacteria to activate sulphate and reduce it to sulphide a number of problems may occur in the offshore oil industry.

In stagnant conditions in oil storage facilities there can be considerable sulphate-reducing bacterial activity and production of hydrogen sulphide which can give rise to two major problems. Firstly, by materials deterioration of the concrete and steel structures on the platform, and secondly by partitioning into the atmosphere, toxic hydrogen sulphide gas can pose a great risk to personnel safety (Wilkinson, 1983).

It has been established that in such environments the sulphate-reducing bacteria are dependent on other organisms both for the development of anaerobic conditions and for the provision of carbon sources in the form of products derived from aerobic crude oil degradation (Ross, 1984; Gilbert et al., 1983).

Inter-relationships between different groups of organisms in oil storage systems have been examined. The enrichment, isolation and characterization of sulphate-reducing bacteria belonging to the genera *Desulfovibrio*, *Desulfobulbus* and *Desulfobacter* are described. In this high sulphate environment the complete oxidation of both lower and higher fatty acids is carried out by the sulphate-reducing bacteria via a two-stage process:

(1)
$$CH_3(CH_2)_{2n}COO^- + \tfrac{n}{2}SO_4^{2-} \rightarrow (n+1)CH_3COO^- + \tfrac{n}{2}HS^- + \tfrac{n}{2}H^+$$
Desulfovibrio sapovorans

(2)
$$CH_3COO^- + SO_4^{2-} \rightarrow 2HCO_3^- + HS^-$$
Desulfobacter postgatei

The enrichment, isolation and characterization of oil-degrading bacteria of the genera *Pseudomonas* and *Micrococcus* was also carried out.

253

Liquid and gel-stabilized model systems can be used in laboratory experimentation on both the temporal and spatial development of oil-degrading communities (including sulphate-reducing bacteria).

In the liquid system the following physical parameters were measured; pH, Eh, $[SO_4^{2-}]$, $[S^{2-}]$ and sulphate-reduction rate. The pH, Eh and $[SO_4^{2-}]$ were shown to fall with time whereas the $[S^{2-}]$ and the sulphate-reduction rate were shown to increase with time. Enumeration of oil degraders, total aerobes, total anaerobes and sulphate-reducing bacteria was also carried out. Initial increases in the numbers of oil degraders, total aerobes and anaerobes were followed after some days by increases in sulphate-reducing bacteria. Increases in numbers of sulphate reducers correlate well with the fall in both Eh and $[SO_4^{2-}]$ and the increase in $[S^{2-}]$ and sulphate-reduction rate.

The gel-stabilized models showed both the temporal and spatial development of oil-degrading communities, oil-degrading organisms being found in the upper few centimetres of the gels and sulphate-reducing bacteria being confined to the bottom of the gels. Gels in which the sulphate was omitted or which contained molybdate, a specific inhibitor of sulphate-reducing bacteria, showed no development of sulphate-reducing bacteria. Similar results have been demonstrated previously (Wimpenny *et al.*, 1981).

REFERENCES

Gilbert, P. D., Steele, A. D., Morgan, T. D. B. and Herbert, B. N. (1983). 'Concrete Corrosion', in *Microbial Problems and Corrosion in Oil Product Storage*, (ed. E. C. Hill), Institute of Petroleum.

Ross, D. (1983). 'Ecological Studies on Sulphate-Reducing Bacteria in Offshore Oil Storage Systems', *PhD Thesis*, University of Aberdeen.

Wilkinson, T. G. (1983). 'Offshore Monitoring', in *Microbial Corrosion*, The Metals Society, London.

Wimpenny, J. W. T., Coombs, J. P., Lovitt, R. W. and Whittaker, S. G. (1981). 'A gel-stabilised model ecosystem for investigating microbial in spatially ordered solute gradients', *Journal of General Microbiology*, **127**, 277–287.

METHODS OF EVALUATION OF BIOCIDE PERFORMANCE IN OILFIELD APPLICATIONS

R. S. Tanner, T. K. Haack,
R. F. Semet, D. A. Shaw and D. E. Greenley

Rohm and Haas Co., Spring House, Pennsylvania, USA 19477

The performance of biocides in the oilfield has been examined in a variety of applications. Emphasis has been placed on developing realistic laboratory models and the measurement of fluid properties. Three different areas are starch-based drilling muds, guar-gum stimulation fluids and bacterial fouling in water injection systems.

Starch Based Drilling Muds

Microbial degradation results in greater fluid loss, reduction in viscosity and corrosion associated with high bacterial populations. In tests designed to monitor these properties a biocide treatment has been shown to protect drilling muds against microbial degradation over a four-week period.

Guar-gum Based Stimulation Fluids

Stimulation fluids are well known to suffer from microbial degradation resulting in loss of viscosity and cross-linkability of the guar. The ability of a range of biocides to prevent such degradation has been evaluated and is demonstrated with regard to these properties.

Water Injection Systems

It is now accepted that biofouling in water injection systems leads to corrosion of equipment and injection of high numbers of organisms into the formation. A dynamic test loop has been designed to realistically evaluate biocides for water injection. Biocides are evaluated under anaerobic flowing conditions. Sampling methods and enumeration techniques have been greatly advanced. Results of experiments designed to determine optimal biocide dosing strategies show that treatments can be designed which cost-effectively inhibit the growth of microbial biofilms.

BIOCIDE TESTING FOR THE NORTH SEA OIL INDUSTRY

P. N. Green, I. J. Bousfield and A. Stones

National Collection of Industrial & Marine Bacteria Ltd., Torry Research Station, P. O. Box 31, 135 Abbey Road, Aberdeen, AB9 8DG, Scotland

The way in which microbial populations develop on submerged surfaces is becoming the subject of increasing attention. Under suitable conditions, clean surfaces rapidly become coated with a complex, often strongly adherent 'biofilm' consisting of micro-organisms embedded in a matrix of organic and inorganic material. A relevant example is the microfouling of water injection systems in the North Sea oil industry by sulphate-reducing (SRB) and other bacteria.

Injection systems are commonly treated with biocidal chemicals in an attempt to inhibit the development of such biofilms, but the results of such treatments are often far from satisfactory. This is often due to the fact that application chemicals have been tested only against the planktonic (free swimming) bacteria and not the sessile (embedded) organisms present in a biofilm. Several laboratory studies have shown that organisms embedded in a complex biofilm are often protected from the lethal effects of various chemical agents largely because these agents penetrate the outer layers of the matrix only to a limited extent.

In an attempt to develop a more meaningful method for the evaluation of biocides against sessile populations Costerton and his colleagues in Canada and the USA devised a laboratory rig for the generation of biofilms — the Robbins device. The laboratory generation of biofilms has been demonstrated on a series of modified Robbins devices built by NCIMB Ltd and used to assess their usefulness in biocide evaluation.

OPTIMIZING BIOCIDE REGIMES IN INDUSTRIAL SYSTEMS USING LABORATORY AND FIELD TRIALS

Peter F. Sanders, Margaret E. J. Barr and David M. Holt

Micran Ltd., Berryden Business Centre, Aberdeen, AB2 3SA

Simple laboratory test devices ('biofilm generators') have been developed which allow screening of biocidal agents against sessile bacteria isolated from pipewalls of industrial plant, in this case the water injection system of a North Sea oil production platform. Bacteria freshly isolated from the system are encouraged to grow on the surface of steel studs in the biofilm generator to simulate sessile bacteria in the offshore system. When a stable biofilm has developed, the studs are removed and treated in a stirred test biocide solution for the desired contact time. The stud is then removed and surviving bacteria are enumerated to assess the performance of the treatment. Finally, one selected biocide can be added directly to the biofilm generator to confirm the data from the stirred tests. Although relatively simple, such tests are an improvement on the traditional planktonic tests, allowing a more realistic ranking of chemicals and assessment of the optimum concentration required for microbial control. This concentration is usually very much higher than that indicated from planktonic tests.

Similar studs have been inserted in an offshore system and allowed to foul naturally for approximately three months. These have been used to monitor the performance of the chosen biocide by removing and analysing studs at intervals during and after a biocide slug dose. The data from the laboratory and field trials were in very close agreement, confirming the applicability of simple sessile biocide testing.

When combined, these two techniques have successfully been used to optimize the biocide treatment regime, to identify the build-up of biocide tolerant strains and to select more effective treatments before such strains become predominant. This dual laboratory and field approach has proved highly cost effective and has kept microbial problems (corrosion, sliming, reservoir contamination) in the system to a minimum and has established and maintained a high degree of microbiological control.